THE SHRINKING PLANET: U.S. Information Technology and Sustainable Development

John Elkington
Jonathan Shopley

WORLD RESOURCES INSTITUTE

A Center for Policy Research

WRI Paper #3
June 1988

Kathleen Courrier
Publications Director

Don Strandberg
Marketing Manager

Hyacinth Billings
Production Supervisor

Each World Resources Institute Report represents a timely, scientific treatment of a subject of public concern. WRI takes responsibility for choosing the study topics and guaranteeing its authors and researchers freedom of inquiry. It also solicits and responds to the guidance of advisory panels and expert reviewers. Unless otherwise stated, however, all the interpretation and findings set forth in WRI publications are those of the authors.

CONTENTS

ACKNOWLEDGMENTS

We should like to thank all those in the information technology and information industries, in the remote sensing industry, in national and international governmental organizations, and in some of the leading environmental organizations who shared their thinking on the implications and potential applications of the emerging information and remote sensing technologies. We are particularly grateful to Gus Speth, Jessica Mathews, Peter Thacher, and to Janet Welsh Brown for their constant support through the project. A draft of the report was circulated to most of those mentioned in it, and the comments that resulted were enormously helpful in shaping the final document. Roy Haines-Young and Julia Hailes of John Elkington Associates also helped in many ways. And we owe a special debt of gratitude to Kathleen Courrier for her help in editing and publishing this report.

J.E.
J.S.

FOREWORD

The industrialized countries are now accumulating vast amounts of information about conditions in the developing countries, including that gathered by satellites and stored in computers. High technology allows us to process, transmit, and store prodigious volumes of data that reveal what's happening to the natural resource base in developing countries.

That information is power, and, as John Elkington and Jonathan Shopley's survey of the U.S. information industry reveals, we haven't yet found a way of sharing it with the Third World.

Remote sensing, for instance, is of enormous importance in monitoring environmental changes and planning for sustainable development; yet, its costs are going steadily up. This important analytical tool is not affordable in the countries most threatened by environmental degradation. The United States has privatized the system, while all of EOSAT's Japanese and French competitors are subsidized. The United States still subsidizes the military uses of Landsat data, but not the environmental applications. U.S. policy-makers have not yet recognized the larger national interests at stake in protecting global ecosystems.

In the following pages John Elkington and Jonathan Shopley illustrate the great potential of the U.S. information industry to help meet Third World countries' resource-development challenges. They describe the exciting technological advances afoot in the industry and cite some of the companies whose products and services would help alleviate poverty and protect the resource base—two central, overwhelming problems in the developing countries.

Note too, though, that the company and organizational profiles appended to this report rarely mention Third World activity. A few developing nations—primarily the so-called newly industrialized countries—will be able to buy, adapt, and use these technologies. But for the poorest countries, those with least developed scientific and technical infrastructures, the use of the new technologies may actually widen the gap between rich and poor societies.

As this survey shows, information technology is being used in environmental management in some Third World countries, but the commercial applications remain lamentably small. As great as the potential contribution of these technologies is, the policies to stimulate their use are not in place. U.S. policy contains no incentives to encourage the applications of these technologies to further sustainable development. U.S. business has too little knowledge of how its technologies and services might alleviate development problems in the Third World. And U.S. companies don't get much help acquiring that knowledge or making those applications—and profits. The industry and the rewards are not organized to promote both development goals and profitability, and without policy encouragement, it will not happen.

Ideally, this report will stimulate discussion in the information industry, on Capitol Hill, and in federal agencies on how U.S. industry might contribute to the goal of sustainable development.

The U.S. Agency for International Development has supported limited training and the use of computers. But much more could be done. Agencies interested in Third World development and in the expansion of U.S. exports should be thinking about how policy might support the "niche" markets that the authors discuss and might promote software development in the Third World.

The information industry is but one of many U.S. industries developing and employing emerging technologies. But its members are particularly "solution-oriented" and forward-looking enterprises whose products and services have the potential to address the difficult challenges faced by developing countries. The biotechnology industry is a similarly oriented enterprise whose potential contributions to health and agriculture were highlighted in an earlier Elkington survey, *Double Dividends? U.S. Biotechnology and Third World Development,* published last year by World Resources Institute. Another example, the rapidly evolving waste-reduction and -management industry, will be featured in a third WRI report, which will be available later this year.

We hope that our efforts will raise important questions for business executives, environment leaders, development specialists, and others:

• What major positive contributions can key emerging technologies make to achieving sustainable development and to solving pressing global-scale resource, environmental, and population problems? What might be the negative side-effects or unintended consequences of such technologies?

• What new research policy measures, incentives and disincentives, and new institutional arrangements are needed to speed the availability of emerging technologies with promising payoffs and to guard against undesirable consequences?

• Since both development and environmental needs are most acute in the Third World, what special measures are needed to secure the benefits of emerging technologies and to prevent negative side-effects in the Third World? Can high-tech be "leap-frogged" into the Third World, or will it further separate developed and underdeveloped? Can exports from U.S. high-tech firms to the Third World be increased in an effort to promote sustainable development and U.S. exports simultaneously?

We hope we can help to stimulate a lively discussion around these vital questions.

James Gustave Speth
President
World Resources Institute

I.

INTRODUCTION:
WHEELS FOR THE BRAIN

Information is perhaps the ultimate renewable resource. New technologies are constantly being devised to collect, transmit, store, process, package and display it. As we move deeper into the "Information Age," with a burgeoning share of global GNP produced by the fast emerging information economy, there is growing interest in the role which information technology—and the information industry—might play in the worldwide move toward more sustainable forms of development. Sustainable development was defined by the World Commission on Environment and Development, in its report *Our Common Future*[1], as development which "meets the needs of the present without compromising the ability of future generations to meet their own needs."

In the wake of the Wall Street crash of 1987 came the recognition of the implications of the "globalization" of financial markets, increasingly linked together by computer and telecommunication technologies. But globalization is a process which has been under way for decades, and not simply in the investment world.

It was no coincidence, for example, that *Our Common Future* opened with the words: "In the middle of the 20th century, we saw our planet from space for the first time. Historians," the Commission concluded, "may eventually find that this vision had a greater impact on thought than did the Copernican revolution of the 16th century, which upset the human self-image by revealing that the Earth is not the centre of the universe."

From the images brought back by the Apollo missions to the Band Aid concerts of 1985, we have been made increasingly aware of the fact that we live on a small planet which is rapidly "shrinking." It is shrinking as new transport and telecommunication technologies enable us to travel further and faster, or to send information almost instantaneously to many parts of the globe. But it is also shrinking as human numbers, particularly in Third World countries, and wasteful patterns of consumption, predominantly in First World countries, press ever more heavily on already over-taxed natural resources.

"Over the course of this century," the Commission noted, "the relationship between the human world and the planet that sustains it has undergone a profound change. When the century began, neither human numbers nor technology had the power to radically alter planetary systems. As the

1

century closes, not only do vastly increased human numbers and their activities have that power, but major, unintended changes are occurring in the atmosphere, in soils, in waters, among species and ecosystems, and in the relationships among all these."

When the World Resources Institute (WRI) organized the Global Possible Conference in 1984, it was recognized that these issues could not be successfully addressed without the support and cooperation of business and industry.[2] Those present, including many senior industrialists, recognized that business has a very real stake in the state of the environment. Business depends on adequate resources, social stability, and receptive markets for goods and services, all of which may be undermined by environmental degradation. At the same time, too, it was recognized that large international companies "have the technology and management expertise for successful environmental and resource management and the financial resources for investment and growth."

But it is one thing to call for partnerships among business, government and other interests concerned to promote environmental protection and sustainable development, quite another to build such partnerships. To explore the potential for such partnerships, particularly involving companies active in some of the emerging "solution oriented" technologies, WRI launched its Technology Project. The first report, *Double Dividends? U.S. Biotechnology and Third World Development* (1986), focused on biotechnology's implications for—and potential applications in—the Third World.[3]

The conclusion drawn was that biotechnology would certainly disrupt Third World economies, but could also provide many of the tools needed for sustainable development. Technology alone cannot solve Third World problems, many of which are caused or aggravated by political, cultural, and moral factors. But appropriate technology can be a critical factor in successful development. The point stressed by *Double Dividends?* was that, while it would be a mistake to opt for the "high technology" road to development at the expense of the many traditional technologies which still have tremendous untapped potential for sustainable development, there are a growing number of "appropriate" high technologies.

The same is true of information technology. Like any tool, it can be misused or abused, but also potentially offers means of ensuring that developing countries are supplied with reliable, timely, and usable data on environmental trends, constraints, capabilities, and opportunities. *The Shrinking Planet* focuses on the environmental implications and applications of some of the new information and communication technologies now in use or under development.

Clearly, the nature, scale, and timing of the environmental applications of these technologies will vary considerably among countries depending upon whether they are industrialized, newly industrializing, or predominantly agricultural.

"Any new technology that can be used in environmental research and management can equally well be used to plan and implement environmentally unsustainable development projects."

But, first, what *is* information technology? When the Office of Technology Assessment (OTA) reviewed some of the critical trends and issues facing the United States in the field of information technology (often abbreviated to IT), it distinguished between "the information technology industry (those who make and sell or provide access to communications media) and the information industry (those who use the new technologies to produce and sell new information services and products)."[4]

The OTA report was itself a symptom of growing concern about the economic health of the U.S. information industry, whose lead both in the commercial sphere and, increasingly, in the underpinning science and technology is being challenged by competition from Japan, some of the newly industrializing countries, and the European Community.

In particular, this concern was triggered by Japan's announcement, in 1982, of its "Fifth Generation" computer project. The project represents an attempt to move towards radically new approaches to computer design based on "artificial intelligence" (AI). This term covers a wide range of research designed to expand the capabilities of computers to embrace tasks normally performed by human beings, from speech recognition and translation through to the development of "expert systems," seen as the first commercial product of more than 25 years of AI research.

Assessing the environmental implications of IT trends has proved no easy task. "Modern computer and communications systems," as the OTA itself concluded, "are among the most complex technologies ever assembled by human beings." The dominant trends in information technology, it suggested, have been driven by the introduction of the microelectronic chip—leading to a rapid increase in computer performance and an equally rapid fall in prices. During the fifteen years between 1972 and 1987, the capacity of memory chips rose 250-fold from a maximum of four kilobits to one megabit, while the price of an array of eight 1-megabit chips, providing one megabyte of random-access memory (RAM), has fallen at least 1,000-fold, from more than $150,000 to less than $150.

> *"During the fifteen years between 1972 and 1987, the price of one megabyte of random-access memory (RAM) has fallen at least 1,000-fold from more than $150,000 to $150."*

If, in 1972, someone had predicted such an extraordinarily rapid pace of development, they probably would have been ruled out of court. But this high rate of improvement in microelectronics will continue into the 1990s and probably beyond. "Changes in the information technology products available in the marketplace," the OTA report argued, "will likely be as revolutionary in the next decade as they have been over the last one that has seen the appearance and growth of such technologies as the personal computer, fiber optics, satellite communications, the video cassette recorder, and two-way cable television."

The direct environmental impacts of this activity are unlikely to be very significant. Certainly IT companies have caused river and groundwater pollution problems, while some of the solvents used to clean chips have been implicated in the destruction of stratospheric ozone. Recent research has also suggested that miscarriage rates may be higher among workers in the "clean rooms" where chips are assembled than in the population at large.

Clearly, too, any new technology that can be used in environmental research and management can equally well be used to plan and implement environmentally unsustainable development projects. Satellite remote sensing is a case in point. While there are a growing number of environmental applications for the information produced by such "eyes in the sky," the main commercial users of satellite remote sensing data to date have been corporations or governments wanting to pinpoint and exploit mineral and other natural resources. The U.S. Agency for International Development has had a remote sensing program since 1971, for example, but, as one slightly jaundiced AID official observed: "There often would seem to be a direct correlation between the remote sensing capabilities of a country and the rate at which it destroys its natural resource base."

The policy context in which such technologies are used is obviously critical. But it is worth stressing that remote sensing—like other forms of IT—has found many positive applications in the environmental field, and has been widely used in attempts to predict and control natural disasters. Remote sensing data contributes to the Famine Early Warning (FEW) program supported by USAID and reported to the UN Office of Emergency Operations in Africa and data have helped provide an early warning system for desert locusts, which threaten an estimated 30 million square kilometers covering many developing countries. By detecting and mapping locust habitats, and monitoring seasonal variations in the soil moisture and vegetation which can trigger locust swarms, effort can be diverted from broad focus aerial and ground surveys to intensive control operations in heavily infested areas.

Such technologies will be used, however, only if they are cost-effective. As Chapter III explains, the privatization of EOSAT and the emergence of the French private sector contender, SPOT Image, have increased the price of remote sensing data. As a result, many developing countries have found it harder to justify the expense even for "mainstream" applications, let alone for environmental applications.

Nevertheless, the strong overall downward pressure on IT prices will help open up totally new applications for computers and other systems. Steven Jobs, who co-founded Apple Computer (page 55) ten years ago, talked of putting a computer into everyone's hands as a way of transforming the world in a million unpredictable ways. But what, we should ask, are the environmental implications of the growth of computerization and computer "literacy" likely to be?

Apple has described its computers as "wheels for the brain." Their environmental impact, and the impact of other forms of IT, will to a large extent depend on the objectives and abilities of those who use them. For example, the self-same computer used by the Department of Defense to launch and target intercontinental ballistic missiles could, in the hands of "nuclear winter" researchers, help model the impacts of the resulting explosions on world climate—or help develop strategies for defusing some of the tensions which could lead to nuclear exchanges. But many key areas of information technology R&D are still predominantly driven, as innovation has been throughout the ages, by military requirements and budgets. The OTA estimated that nearly 80 percent of Federal IT funding comes from the Department of Defense.

"The same computer used by the Department of Defense to launch and target intercontinental ballistic missiles could, in the hands of 'nuclear winter' researchers, help model the impacts of the resulting explosions on world climate."

Presumably, however, since computers are far from infallible, it is only a matter of time before computer failure or "human error" embodied in software precipitates a major military or industrial disaster. The experience of computer failures during the NASA Space Shuttle missions has shown how vulnerable such complex systems can be, even when computer components are duplicated or triplicated. And there are likely to be wider implications of computerization. The use of computerized stock-dealing systems on the Wall Street, London, and Tokyo

stock markets, for example, undoubtedly helped aggravate the worldwide collapse of stock prices in late 1987.

But the fact that a certain proportion of cars or aircraft crash every year does not stop people driving or flying, and the momentum of computerization is now such that ultimately this process will touch almost every part of our lives. Computers not only enable us to store extraordinary volumes of data, but increasingly also help us to "think the unthinkable." Some of the problems and opportunities now emerging in the field of sustainable development will require both abilities.

"So far, the IT industry has tended to back into the markets being created by the growing demand for environmental protection and sustainable development."

So far, the IT industry has tended to back into the markets being created by the growing demand for environmental protection and sustainable development. Increasingly, however, U.S. computer companies—because they continue to dominate the world IT market and some of their products are inevitably used in environmental research or management applications—also find that they are becoming involved in some of these new market niches.

IBM, which has been going through a difficult period, has been looking for future market niches and, as Chapter V explains, has considered both remote sensing and environmental management as possible large-scale consumers of computing power. The U.S. industry, in fact, produces the world's most powerful computers, some of which have found environmental applications. When Britain's Science and Engineering Research Council wanted a supercomputer able to store and process the prodigious volumes of data needed in analyzing the circulation of the atmosphere and oceans, for example, it chose a Cray X-MP/48.

In recent years, however, exports of such systems have increasingly fallen foul of U.S. controls on so-called "dual-use" technologies, commercial products which could be used for military purposes by hostile powers. Considerable controversy ensued when India wanted to buy a Cray machine for modelling monsoons, to help predict and minimize the impact of the resulting flood damage. Concern was expressed that such a supercomputer could be used in developing a nuclear strike capability, or that the machine's internal "architecture" might fall into the wrong hands. It is worth noting, however, that even an automated banking machine may contain information-processing chips which could be used in missile-guidance systems.

At a time when the computer industry has been reporting lower growth rates, as described in Chapter II, such constraints have caused significant problems. Clearly, companies involved in the information economy are in business to make money. Within broad limits, they will go where the markets are—or are likely to be.

If a company like Corning Glass (see page 11), to take just one example, were to find that its new line of transparent cookware products significantly and consistently outperformed its optical fiber products in terms of financial returns, it might well take the view that it should be in cookware rather than telecommunications. In that sense, at least, information technologists are no more likely to represent a natural constituency supporting sustainable development than any other cross-section of industrialists.

"Information technologists are no more likely to represent a natural constituency supporting sustainable development than any other cross-section of industrialists."

But some of the companies profiled and individuals mentioned in the following chapters *have* been thinking about the market opportunities likely to be opened up by sustainable development. This convergence in commercial and environmental objectives should be recognized, where it occurs, and exploited. Some of the issues for policy-makers, particularly those relating to Third World development, are highlighted in Chapter VI.

Third World countries have often experienced considerable problems in absorbing new technologies, including environmental technologies. As a result, advocates of "appropriate" technology have often tended to favour low-tech solutions in Third World problems. There is at least a possibility, however, that some of the trends now emerging in IT could result in appropriate high-tech solutions.

This argument has long been advanced in the space communications field. At the first U.N. space conference in 1968, Dr Vikram Sarabhai suggested that Third World countries could well "leapfrog" the West by using satellite communication systems from the start, rather than investing in already obsolete conventional telecommunication systems.[5] What chance is there that the Third World might be able to "leapfrog" the First and Second Worlds in the broad area of IT? A growing number of Third World specialists think that there is a good chance, with countries like Brazil and India banning imported minicomputers and personal computers in an attempt to boost domestic manufacturers.[6]

"With 70 percent of the world's population, the Third World has barely 7 percent of the world's telephones."

But an indication of the extent to which the Third World as a whole lags behind the industrialized nations is provided by the statistics on telephone ownership: with 70 percent of the world's population, the Third World has barely 7 percent of the world's telephones. "On average," as *South* magazine reported, "industrialised nations have 50 telephones per 100 persons; the less industrialised nations have only 3 per 100. There are more telephones in Tokyo than on the whole continent of Africa."[7] Africa, in fact, has 0.7 telephones per 100 Africans, one-hundredth the ratio found in the United States, and 30–40 percent of these telephones are estimated to be out of service at any one time.

These sobering facts should be borne in mind when considering the undoubted potential of some of these new technologies. "There is no technical limit to what can be done," as one image enhancement consultant put it, an observation which applies to most of the areas touched on in this report. But, as Robert Noyce, co-inventor of the integrated circuit countered: "A lot of things are technologically possible, but only economically feasible products will become a reality."[8]

The key questions that need to be asked about information and telecommunication technology must include the following: What is likely to happen? What will the implications be? What *needs* to be done, by whom, for whom and at what cost? This short report does not aim to provide all the answers. Instead, it aims to provide some of the information needed to ask the right questions about an industry that looks set to be among the world's largest—if not actually the largest—by the year 2000.

II.

THE I.T. BUSINESS

Until the 1960s, it was impossible to take a picture showing the earth as a whole, let alone generate images covering the whole of the earth's surface every few hours. Today, thanks to satellite remote sensing and the achievements of the information technology (or IT) industry, this is a matter of routine.

The development of the IT industry has not been driven by applications such as the remote sensing of the earth's renewable resources, however. In fact, military and commercial applications have made it one of, if not the, fastest growing industry of the eighties.

"A growing number of Third World countries see investment in information technology as a necessary condition for any attempt to catch up with the industrialized nations."

In 1986, worldwide sales of computer hardware, software, and telecommunications gear were valued at around $350 billion by the Computer & Business Manufacturers Association (CBEMA).[9] And the IT industry is forecast to be the world's largest manufacturing industry by 1990, with a market worth an estimated $1.3 trillion by the turn of the century.[10] The cutting edge of development in this field has been in the United States in recent years, though Japan and a number of other Pacific Rim countries, including Korea and Taiwan, have been making significant commercial inroads. A growing number of Third World countries see investment in information technology as a necessary condition for any attempt to catch up with the industrialized nations. To date, however, even relatively advanced Third World countries, like India and Brazil, have found it difficult to establish a foothold in IT.

The history of modern computing and the IT industry goes back to the mid-nineteenth century, when the English mathematician Charles Babbage designed (but never built) his "analytical engine." The United States took over the leading position in computer research during the 1940s. The early computers were enormous, incorporated massed banks of valves, and required prodigious amounts of electricity to power and cool them. For example, the University of Pennsylvania's ENIAC (Electronic Numerical Integrator and Calculator), built in 1945, contained 18,000 valves and weighed 30 tonnes. But even this heavyweight was not a computer in the full sense of the word, since it had no memory. Machines like these provided the number crunching power to deal with vast amounts of manually collected data.

7

Microchip building blocks

The microchip gave computers the ability to both store and manipulate data. Indeed, the emergence of the IT industry relates closely to developments in microchip technology, which has driven the dramatic reduction in size and cost of computers.

The origins of the microchip industry can be traced back to the invention of the transistor at Bell Telephone Labs in 1947. Raytheon was the first company to exploit transistors commercially, using them in hearing aids, subsequently losing many of its staff to Transitron and Texas Instruments. William Shockley, a co-inventor of the transistor, moved west and founded Shockley Semiconductor in 1955, with the objective of developing silicon transistors. By this stage, UNIVAC had installed the first commercial "mainframe" computer for the Department of the Census in 1951. A further 46 were subsequently bought by business users, at a cost of $500,000 apiece. By 1953, the IBM 701 had appeared, renting for $15,000 a month, and the process of office automation was under way.[11]

The next step in the chain came when eight Shockley employees left to form Fairchild Semiconductor, where Robert Noyce came up with the "chip"—invented simultaneously by Jack Kilby of Texas Instruments. Chips are the integrated circuits that cram thousands of electrical components, including transistors, onto minute pieces of silicon.

Fairchild, in turn, created a further generation of spin-off companies, including National Semiconductor (1967), Noyce's Intel (1968), and Advanced Micro Devices (1969). These new companies accelerated miniaturization, producing ever-cheaper building blocks for new generations of computers. In 1963, another new start-up company called Digital Equipment (see page 61) launched its PDP-8 minicomputer, priced at just $18,000. At last, computers were becoming cheap enough to wait on their users' needs, rather than the other way around.

1977 saw the next milestone in IT history with the launch of the Apple II, followed in 1981 by IBM's best-selling PC. Now the computing power that once required enormous, electricity-hungry machines could be generated by small boxes, or "personal computers" with a "footprint" small enough to sit comfortably on a desk. Whereas major companies once had a mainframe and a few minicomputers, they now had hundreds or thousands of personal computers. By the end of 1986, it was estimated that an electronic keyboard sat on over half of America's 60 million desks and that by 1990 three-quarters of the country's desktops would have a computer screen. Market analysts Dataquest estimated that U.S. business would spend some $116 billion on computers between 1986 and 1990 and $127 billion on the software needed to turn those machines into practical tools.

By 1986, chip manufacture accounted for $250 million in the U.S. industrial output.[12] IBM is the world's leading producer of chips (most of which it uses in-house), with AT&T second. However, such Japanese chip makers as NEC, Hitachi, and Toshiba are now the world's biggest producers for the open market.

The ability of a computer to process information faster and cheaper depends on the amount of information that it can hold in its internal memory, and advances in internal memory storage capacity are continually being announced. While the first personal computers began with 64 kilobytes of internal memory, the frontiers of internal memory are being pushed back by such researchers as Princeton's Richard J. Lipton, who has developed a "massive memory machine" (or M3) that will have access to 1 gigabyte of internal memory.[13]

Commercially available chips can currently store up to a megabit of information, although the next 4-megabit generation of chips has already been developed by IBM, and Toshiba is working on a 16-megabit chip. Companies such as AT&T are working on gallium arsenide chips; faster and more powerful, they also consume less power than silicon chips. At present, memory chips are about ten times more expensive than the slower magnetic disk drives, but Dataquest forecast that they will only be about 66 percent more expensive by 1990. Such trends imply that the size of computers will continue to fall rapidly while their power increases substantially. As chips get smaller and more powerful, however, the possibility of interference between components grows, though the use of 3-D architectures has helped to reduce this problem.

Bigger bytes

The present generation of commercial chips process information in 32-bit chunks, representing a fourfold improvement on the 8-bit chips that started the personal computer race. The U.S. industry has been pinning some of its hopes on the new generation of 32-bit microcomputers, which offer the computing power of a minicomputer at a fraction of the price. For less than $5,000, these machines will provide the computing power that only a few years back would have been possible only with a $900,000 mainframe. At their heart will be chips manufactured by companies such as Intel and Motorola. Compaq was the first to market with an IBM-compatible 32-bit machine, with Apple and IBM among the companies working on rival 32-bit machines.

"The U.S. industry has been pinning some of its hopes on the new generation of 32-bit microcomputers, which offer the computing power of a minicomputer at a fraction of the price. For less than $5,000, these machines will provide the computing power that only a few years back would have been possible only with a $900,000 mainframe."

The 32-bit workstations produced by such companies as Apollo Computer, Silicon Graphics, or Sun Microsystems sold for less than 20 percent of the price of a DEC minicomputer by 1986, and prices will fall further as mainstream competitors, including DEC, IBM, Hewlett-Packard, NEC, and Apple swing into action. Such systems will be much more "user-friendly." Indeed, they have so much computing power that they can devote 75 percent of their time to running software that helps the user interact with the machine, yet still run through their tasks faster than ordinary 8-bit or 16-bit PCs. From a Third World point of view, the rapid pace of technological innovation may be confusing, but longer term these trends are important because they promise cheaper, easier-to-use computers.

"Anything a minicomputer does now," said Compaq's Vice-President of Marketing, Michael S. Swavely, "a micro will do in two years."[14] Eventually, even the largest and most sophisticated computers could well face competition from 32-bit machines, with "parallel processing" (using many microprocessors simultaneously to deal with various parts of a single computing job) seen as a way of achieving greater computer power at lower cost. Masscomp is one company that claims it will use this "multiprocessor" approach to build a computer that could challenge the power of a $5-million Cray supercomputer by 1989 or 1990.

IBM scientists Gerd Binnig and Heinrich Rohrer won the 1986 Nobel Prize for their development of the "scanning-tunnelling microscope," which can map the position of individual atoms on a surface. Bell Laboratories have used the instrument to control the deposition of individual atoms of tungsten onto semiconductor material and have demonstrated that they can be read back as information. This could be the first step toward the next generation of extremely small, but very fast memory chips.[15]

Cheaper computing

Such chips will help build new supercomputers, more powerful than those currently offered by such companies as Cray (see page 60) and ETA Systems (page 59).

In this field, too, the Japanese are forging ahead, while research elsewhere suggests that the cost of supercomputing could fall significantly. The British microchip manufacturers Inmos, a subsidiary of Thorn EMI, for example, have developed the "transputer," a computer on a chip. Southampton University, supported by the European Community's Esprit Programme, linked 320 transputers together to create a supercomputer that rivalled Cray's machines but costs only $750,000, compared to several million dollars for a Cray.[16] If the costs of supercomputing can be brought down, it could be used much more widely than in the defense and R&D laboratory applications it has found to date.

From 1972 to 1987, the capacity of memory chips increased 250-fold, while the price of one megabyte of random-access memory (RAM) was reduced by at least 1,000-fold.[17] Meanwhile, just as chips eventually became an international commodity, increasingly sold at commodity prices, so too have certain types of computers. The implication is that once basic standards are set, computer production will become a business like any other, with competition in maturing markets increasingly focusing on price.

However, although many computer and IT companies continue to show rapid growth and there continues to be a proliferation of new IT products, 1985 saw something of a watershed for the U.S. computer industry. Instead of the extraordinary annual sales growth rates to which it had become accustomed, the industry sold only 10 percent more computers in 1985 than it had in 1984—only 4 percent more if personal computers were excluded from the equation. Dataquest predicted that the growth in the U.S. computer industry's total domestic and overseas sales would rebound during 1987, but it was among those arguing that future growth in sales to American business would stay significantly lower, predicting an average annual growth rate of just 9 percent up to 1990. Among the factors driving these trends are falling prices, saturated markets and the growing emphasis on new ways of linking up computers already in use—rather than on buying new machines.

Hardest hit have been sales of big mainframes, with low growth rates expected through 1990. This is a worrying trend for IBM (see pages 37 and 64), which holds about two thirds of the global mainframe market. With total sales of more than $50 billion, IBM is described by other major mainframe computer companies not so much as the competition, but as "the environment." Even Unisys, formed when Burroughs and Sperry merged, accounts for a total share of the mainframe market that is only about one-sixth of IBM's. Some U.S. companies believe they can compete with IBM only by linking up with foreign companies: Honeywell, for example, sold 57.5 percent of its computer business to Japan's NEC and France's Groupe Bull, while Japan's largest computer maker, Fujitsu, has invested heavily in Amdahl.

As for minicomputers, IBM's main competitor has been Digital Equipment (DEC). But DEC itself is under pressure from smaller rivals such as Data General. The personal computer market, meanwhile, has been showing dramatically lower annual growth rates when compared with the 40 to 100 percent-plus growth rates it once enjoyed. In 1986, U.S. personal computer sales grew by just 5 percent.[18] Here, too, IBM dominates the market, although there is

competition from companies like Compaq and AT&T, which have also entered the IBM-compatible market. Apple Computer (see page 55), still strong in the educational and home markets, has been gradually building up its share of the business computer market—particularly with its Macintosh-based desktop publishing systems. It is also now linking with DEC to compete more effectively with IBM.[19]

Asian producers of "clones," cheaper computers able to run IBM software, are putting pressure on one market sector after another. U.S. manufacturers may experience even more problems as computer prices are driven down and as cheap Asian chips create a magnet for computer industry investment in the region. But the range of computer applications will presumably expand rapidly as the price of computing power continues to fall. Further joint ventures are being established between U.S. and Japanese companies to exploit fast-growing Asian markets. Motorola and Toshiba, for example, announced a technology exchange agreement late in 1986 and are aiming to build a jointly-owned chip plant in Japan. At the same time, too, such countries as South Korea have shown that they can also innovate. A case in point is Golden Semiconductors' new version of the 256 SRAM (static random access memory) chip, which operates at twice the speed achieved by its main competitors, made by Toshiba and Hitachi of Japan. Although SRAM's are most likely to be used in advanced weaponry systems, they are also used in telecommunications equipment.

Computer networks

Instead of trying to sell more computers than IBM does, companies like DEC are now developing and selling computer "networking" systems, which link different types and makes of computer. Ken Olsen, DEC's founder and president, sees networks as the business of the 1990s. "Network technology is one of the most complex in the world," he explained, "and once a network is in a company it can die without it." Even Apple, which has consistently resisted IBM-compatibility, is now offering such systems as Appleshare, a relatively simple and inexpensive system that enables Apple and IBM machines to communicate.

With the power of computers increasing as it has, individual machines are now able to perform tasks which are of more than local importance. In response to the need to connect individual computers into networks, the IT industry is currently focusing much of its R&D effort on computer "connectivity." Digitally stored information is increasing by 40 to 50 percent a year, and the need to exchange data between different systems is growing in parallel.

> "The telecommunications industry is currently focusing much of its R&D effort on computer 'connectivity.' Digitally stored information is increasing by 40 to 50 percent a year, and the need to exchange data between different systems is growing in parallel."

The computer industry has developed Local Area Networks (LANs), which permit different computers to intercommunicate. At the Battelle Memorial Institute's Pacific Northwest Lab, for example, LANs have provided a structure within which a Cray supercomputer, three IBM mainframes, a UNIVAC mainframe, eight large DEC minicomputers, 50 MicroVax terminals, and over 2,500 other desk-top computers can inter-communicate. Local Area Networks (LANs), have taken a long time to break through in the marketplace, but the entry of IBM into the field, initially with PC-NET and then with Token Ring, has greatly increased both user acceptance and competitive pressure. "Networks will open up the 80 percent of the business market that hasn't been reached by personal computers" says Charles W. Berger, vice-president for marketing at Apple. Further, networks are being used by companies like Excelan to overcome incompatibility problems which have previously prevented the linking up of dissimilar computers.[20]

Longer term, the development of the Integrated Services Digital Network (ISDN) holds enormous promise for data communications. ISDN will be digital throughout, carrying voice as well as data and video channels. "The future," as AT&T Vice-President Tom Kerr explained, "lies in the evolution of today's mixed analog and digital networks into an [ISDN]. As customer needs expand, so will ISDN, until it evolves into a network that can provide what we call universal information services."[21] Or, as William G. McGowan, Chairman of MCI Communications, explained, "We're combining computing power, information technology, and telecommunications into a transportation system for the information age."[22]

A consequence of this expanding market opportunity is that computer companies are likely to find themselves competing with telecommunications companies in the race to capture networking markets. This competition is already beginning to spill over into Third World markets, like India, where such companies as Alcatel, Ericcson, ITT and Siemens are competing for what some see as a potential $2 billion annual market in technical expertise and equipment sales.

> "Such companies as Alcatel, Ericcson, ITT and Siemens are competing for what some see as a potential $2 billion annual market in technical expertise and equipment sales in developing countries."

Fiber optics: riding the lightwave

As far as regional and worldwide data communication is concerned, the breakup of AT&T (see page 56) in 1984 has stimulated U.S. investment in advanced communications projects. In the home market, the basic goal has been the "rewiring of America," with enormous investments now going into work designed to create an eventual worldwide, homogenous telecommunications network running entirely on digital equipment.

The pace of development has been spurred by the emergence of fiber optic technology. The possibility that pulses of light might be used to carry telecommunications traffic rather than the conventional electrical pulses was suggested when Corning Glass produced a glass fiber as thin as a human hair in 1970. To transmit conversation, a fiber optic system converts voice signals into digital "on" and "off" code, which is then pulsed down the optical fiber by a laser—with almost no distortion or information

11

loss. The transparency of silica glass fiber has now approached the theoretical maximum, while enormous progress has also been made with the "plumbing" (amplifiers, switches, couplers, filters, and isolators) needed to carry the light signals.[23]

Japan, which spent an estimated $3 billion on optoelectronics in 1985, leads the field, but there is strong competition not only from the United States (which spent at least $1 billion that year), but also from Europe and the Soviet Union (estimated to have spent four to ten times as much).[24] Among the growing number of U.S. companies active in optical technology R&D are AT&T, Du Pont, GTE, IBM, 3M, NCR, and Texas Instruments.

The value of optical fiber is illustrated by the difference between the performance of early undersea telephone cables and their modern optical fiber replacements. The first undersea cable between England and the European mainland, laid in 1853, consisted of four copper wires insulated by natural rubber and carried four telegraph signals. In 1986, UK-Belgium 5 was laid, consisting of three optical fibers carrying just under 4,000 simultaneous telephone conversations. Ultimately, it is theoretically possible to carry the entire 1985 telephone voice traffic in the United States on a single fiber. Among the U.S. companies actually laying fiber optic networks are AT&T, Sprint (a joint venture between GTE and United Telecommunications), and MCI.

The first trans-Atlantic fiber optic communications cable (TAT-8), due for completion in 1988, is designed to link the United States with Europe and poses a possible threat to the satellites that currently handle 60 percent of transatlantic phone calls. With transatlantic telephone traffic expected to grow at 20 percent a year, coupled with the ability of such a fiber optic link to handle 40,000 simultaneous conversations (equivalent to the capacity of the advanced INTELSAT VI satellite), the attractions of the technology are clear. AT&T has a 37-percent stake in the venture and will build the first 3,000 nautical miles.

According to John S. Mayo, Executive Vice President of Network Systems for AT&T Bell Laboratories, the photonics—or "lightwave"—revolution has barely begun, though its impact will eventually dwarf that of microelectronics technology. "We probably have two more decades of progress in lightwave before we reach the ultimate physical limit as we now understand it," he said, predicting a thousandfold increase in transmission capability over today's most advanced research result of 20 billion bits of information per second through a single optical fiber.

It may be some time before Africa and South America are integrated into global telecommunications networks by fiber optic links. However, Third World countries everywhere may benefit from advances in satellite communication technology driven by the competition between the satellite and optical fiber industries.

The optical challenge is certainly spurring further developments in satellite technology. The use of lasers could increase a satellite's capacity tenfold, with no increase in weight. MIT's Lincoln Laboratory is developing a prototype space laser communications system known as the Lasercom Package, with backing from the Strategic Defense Initiative's Innovative Technology Office. The end-result could be a network of laser-connected satellites able to communicate from any point on earth to any other without the need to link through ground-relay stations.[25]

"The optics boom is just starting to explode," as Robert Spinrad, Director of Systems Technology at Xerox, put it.[26] Optical technology will dramatically affect the capacity of telecommunications links; already it is having an impact on many sectors of the information business. Besides the challenge that optical fibers pose to conventional copper and satellite communication networks, research is now under way on new generations of "optical computers" that will use laser pulses rather than electrical currents to carry information. An area where optical technology is already well entrenched is in the permanent storage of computer data.

The data storage challenge

The age of the computer has seen an exponential growth in data—and these data have to be stored before and after computer processing. Present methods of data storage are largely based on silicon memory chips, or on such magnetic media as floppy disks or winchester disks. Floppy disks have storage limitations, while memory chips and winchester disks are often computer-bound, though

LANs (and, ultimately, ISDN) promise to unlock data held on such computers for a network of users. Further, although the capacity of memory chips has increased dramatically, they are generally employed to "hold" data during computation, rather than to store data long term.

The same 4.7-inch "compact" plastic disks and optical disks that have revolutionized audio recording are now being developed for computer data storage applications. A single "compact disk read-only memory" (CD-ROM) can hold 550 megabytes of data, equal to 1,500 floppy disks or more than 200,000 pages of text—enough, as the magazine High Technology put it, to hold "the Encyclopaedia Britannica or [the] complete SEC data on every company listed on the New York and American Stock Exchanges."[27] The data are encoded onto a plastic disk with a laser beam, a process that can only be done once, and can then be read off innumerable times by a scanning laser beam.

"By 1987, data storage on a CD disk cost 17.5 cents per megabyte compared to $5 per megabyte for equivalent floppy disk storage. A single CD disk can store roughly the same amount of data as 1,500 floppies, and the cost of CD disk drives is also falling to increasingly competitive levels."

These CD or optical disks are much more robust than floppy disks, and their price has been tumbling. By 1987, data storage on an optical disk cost 17.5 cents per megabyte compared to $5 per megabyte for equivalent floppy disk storage. A single optical disk can store roughly the same amount of data as 1,500 floppies, and the cost of optical drives is also falling to increasingly competitive levels.[28]

Eastman Kodak has assembled 150 optical disks into a "data jukebox" which provides a storage capacity of more than a tetrabyte (1,000 gigabytes) of data. The system is targeted for use in hospitals where it will be able to store more than 12 years worth of X-rays and medical records for a 250-bed hospital and access any of that information within 12 seconds.[29]

The major disadvantage of CD disks is that they are an order of magnitude slower than magnetic disks. But market analysts Frost & Sullivan see the market for optical disk media, and the associated hardware, growing from a 1986 value of $90 million to $2.5 billion by 1991.[30] Although the market has been driven by music rather than data storage applications, the emergence of this technology means that large databases previously held on enormous computers, and accessible only through such online services as Dialog and The Source, are increasingly available on optical disks. The disks can be sent to individual users through the post.

IBM and Matsushita Electric Industrial Company of Japan are locked in a race to overcome the other great disadvantage of optical disks, which is that they can only be encoded with data once. Both companies, and others too, like Eastman Kodak and Sony, are working on the next generation of optical disks—which can be erased and rewritten.[31]

Sensors: harvesting the data

Clearly, the IT industry is developing an impressive array of products to store, transmit, and process very large volumes of data. This ability to "consume" data has been paralleled by the evolution of a sophisticated array of digital data gatherers, or sensors. In fact, advances in sensor technology have effectively opened the digital data floodgates to the point where the next generation of IT products are once again being driven by the unmet demand for data handling power.

In 1978, for example, NASA's Seasat satellite carried an experimental Synthetic Aperture Radar (SAR). For the three months that the sensor was operational, it returned more data about surface winds over the oceans than had been collected in the previous 100 years from all monitoring sources.[32] This single example clearly illustrates the potential contribution that remote sensing technologies could make in providing baseline and monitoring data to those developing countries whose efforts in sustainable development are hampered by inadequate environmental resource data.

The U.S. Landsat remote sensing program was initiated in 1972 with the launch of Landsat 1. Onboard was the Multispectral Scanner (MSS) sensor, which measures the energy reflected by the earth's

13

surface. Orbiting the earth at an altitude of about 570 miles a Landsat platform circles the earth 14 times a day, producing a complete set of images of the whole globe in all four original wavebands every 18 days.[33]

Sensor technology has evolved rapidly, with improvements in spectral resolution (by adding more wavebands), temporal resolution (by scanning the globe more frequently), and spatial resolution (by using smaller pixels). Landsat 5, launched in 1984, carried the Thematic Mapper (TM) sensor in addition to the MSS. The TM has a ground resolution of 90 by 90 feet, together with additional wavebands. Other countries such as France, Japan, and the European Community have incorporated remote sensing into their own space programs, and have also developed more refined sensors.

For example, active microwave sensors (radar) have the ability to "see" through cloud and darkness. Unlike the MSS or TM scanners, they emit their own radiation and measure the returning reflections. Side-looking radar has an added advantage: it can highlight subtle structural features that would be masked by vegetation in aerial photographs or Landsat imagery. The ability to penetrate cloud and vegetation cover offers some hope for those countries in tropical and equatorial regions whose resource inventories have been severely restricted by such environmental factors. The success of the Seasat SAR sensor has ensured that similar sensors will be on board the European Space Agency's ERS-1 satellite, as well as the U.S. Navy's Remote Sensing System (NROSS).

While satellite-borne sensors provide a synoptic view of the global environment, fiber optic sensors, which can function in extremely harsh environmental conditions, are finding new applications inside nuclear reactors, turbines, petroleum and chemical processing plants, and other industrial environments characterized by extreme temperatures, highly flammable materials, toxicity, and other unstable conditions. These applications will result in "superfast" growth through the 1990s, according to market analysts IRD.[34]

Biosensors, which combine biological membranes or cells with microelectronic circuitry, also promise to evolve into highly specific sensors, able to detect miniscule quantities of pesticides or other potentially hazardous materials. Japanese researchers are in the lead in the new hybrid field of bioelectronics. Five government ministries and more than 50 Japanese companies have entered the field in the last few years, and some analysts predict a billion dollar market by 1995. Companies such as NEC and Mitsubishi Electric aim to shrink biosensors to the size of a microchip. Professor Isao Karube of the Tokyo Institute of Technology has been working on yeasts extracted from waste-water sludge to develop biosensors for detecting organic pollution in water. He has also been working on other systems for detecting "acid rain" and other forms of air pollution. Ultimately, computer circuits could be built into such biosensors, enabling them to analyze the data they collect on the spot.

> *"Japanese researchers are in the lead in the new hybrid field of bioelectronics. Five government ministries and more than 50 Japanese companies have entered the field in the last few years, and some analysts predict a billion dollar market by 1995."*

The Japanese believe that, when mass produced like microchips, such biosensors could retail for the equivalent of around 60 cents, compared with prices several hundred times greater for today's early models. Although several American companies were the first to market biosensors, they have since fallen behind—both in process control and potential environmental applications. Brewers like Asahi may have had problems in keeping their yeast-based alcohol sensors alive for more than a few days, but overall the evidence suggests that biosensor research is going to produce commercially viable sensors for a wide range of applications. Although the United States, Britain, and France are strong in such research, Japan seems to be gaining a major competitive edge.[35]

Software: pulling it all together

Clearly, the technologies for the collection, transmission, processing, and storage of data have become progressively more powerful and affordable. But what about data management? The importance of computer software, of the programs that help transform data into information, is growing rapidly. As far as the actual and potential environmental applications of IT are concerned, these fall into a number of major categories, including image processing, geographical information systems and, longer term, artificial intelligence. These three application areas are discussed below.

The Jet Propulsion Laboratory began applying digital computers to *image processing* for NASA's Ranger and Mariner missions in 1963, with IBM 7094 computers used to assemble pictures from Mars by 1965. Later, with the advent of earth observation multispectral scanner sensors, the Laboratory for Applications of Remote Sensing (LARS) at Purdue University developed the first interactive digital display for image processing. With the advent of the Landsat satellite series, more advanced methods were needed. The Mariner video data were sent back at the rate of 8.33 bits per second, while early Landsat thematic mapper data came in at rates of up to 85 million bits per second.

By 1985, the power of the IBM PC was harnessed by software developed by Myers and Bernstein at IBM's Palo Alto Scientific Center (page 65). The combination almost matched the image-processing abilities of the original IBM 7094, and the cost of the equipment was now down to around $10,000.[36] With an average-sized image consisting of 256,000 bits of information (or pixels), 32 kilobytes of computer memory would be needed to store a single image. Over the past fifteen years, however, the capacity of computer memory chips has risen from 4 kilobits to 1 megabit, and is being continuously expanded.

Early commercial applications, however, are just as likely to be in totally unrelated areas, such as the film industry, where George Lucas used one of the first super-workstations to generate images for his "Star Wars" films. Pixar, which Steven Jobs joined on leaving Apple, has developed integrated image processing systems of use in hospitals, computer-aided design, and the analysis of seismic survey results. Other innovators include Elizabeth Arden (whose "Elizabeth" system simulates the application of make-up to a client's face, hardly a priority in the Third World) and Information Builders (whose 3-D simulations of houses allow online house buying). But these diverse applications are symptoms of an underlying trend. "The time has come," as Pixar Vice President Alvy Ray Smith put it, "when technology is cheap enough that people can start affording instruments they couldn't touch before."

The increasing availability of digital image data is helping spur the development of new *geographic information systems* (GIS). A GIS is a computerized database incorporating elements of cartography, geography, photogrammetry, remote sensing, statistics, surveying, and many other disciplines concerned with the analysis of spatially-referenced data. (See Appendix for profiles; Autometric—page 57, ERDAS—page 63, and ESRI—page 63).

An example of a global GIS is UNEP's Global Resource Information Database (GRID), inaugurated in 1985 and due to become operational in 1988. The GRID computers, based in Nairobi and Geneva, will help open up the data collected by GEMS, the Global Environment Monitoring System. Three software systems are being used by GRID: ELAS, developed by NASA's Earth Resources Laboratory, implemented on a Perkins-Elmer minicomputer; ARC/INFO, developed by ESRI, which runs on a Prime computer; and ERDAS image processing software, loaded on an IBM PC/AT microcomputer. Developments in software engineering and the growing power of microcomputers have enabled ESRI to produce a version of its ARC/INFO system able to run on an IBM PC/AT. All these advances combine to provide GRID with the means to incorporate almost any dataset into its system, and once entered, to analyse any combination of environmental parameters. This also gives the GIS system an enormous capability to run "what if?" scenarios involving changed conditions in any of the datasets.

In 1983, ESRI coordinated a conference for NASA on the prospects for large-scale GIS development, and the conclusions remain topical. One of the major cost elements in GIS work, it was noted, is the inputing of data. Thus, automation of this stage remains a priority. It was also concluded that software was more of a limiting factor than hardware, so the need to increase GIS "user-friendliness" is pressing. Once the data are on a GIS, there is a need for "data browsing" methods, which "expert systems," discussed below, could well help advance. Developments like these are being pursued by research and training institutions (ESRI and ERIM among them), and commercial companies like Autometric.

The potential contributions of geographical information systems to environmental management and sustainable development in Third World countries is clearly very considerable, although there are significant constraints on their ability to exploit such technology. Given the key role of software in harnessing IT hardware in particular application areas, the fact that very few Third World countries have yet built up a software development capability is worrying. And the gap between the developed and developing countries seems certain to widen as we move into the era of "expert systems" and "artificial intelligence."

The power of many types of software is likely to increase dramatically with the introduction of early forms of *artificial intelligence* (AI). This, the Office of Technology Assessment reported, "is a term that has historically been applied to a wide variety of research areas that, roughly speaking, are concerned with extending the ability of the computer to do tasks that resemble those performed by human beings. These capabilities include the ability to recognize and translate human speech, to prove the truth of mathematical statements, and to win at chess."[37]

When IntelliCorp was founded in 1980, making it the longest-established company in the commercial AI field, AI applications were confined to specially designed computers out of the reach of most commercial applications. Instead, most applications were defense-related R&D programs. Two AI programing languages emerged as standards—LISP (short for "list programing") and PROLOG (short for "programing in logic"). Until recently, AI applications could operate only on AI-dedicated computers built by companies such as DEC, Symbolics, Texas Instruments, and Xerox.[38]

The field really began to take off, however, with the appearance of "expert systems" software based on LISP, but able to run on conventional computers. ExpertIntelligence, initially a developer of educational software, has been selling expert systems for the Apple Macintosh since 1983, while IntelliCorp has developed a $30,000 "shell" system known as Knowledge Engineering Environment (KEE) for use both by AI-dedicated machines and by DEC's Vax minicomputers. The company has also developed PC-Host, which allows expert systems run on minicomputers to be accessed by IBM PCs. Such

software houses as Aion Corporation are now marketing expert systems that run directly on IBM PCs, while Distribution Management Systems have gone even further, abandoning AI languages and opting for COBOL, a popular "fourth generation" language that is easier to embed in commercial programs.[39]

This shift in the expert system market wrong-footed AI companies like Symbolics, which had captured half the 1986 market for specialized AI computers. Texas Instruments, another leader in this field, sidestepped the problem by coming out with its LISP-on-a-chip, which packs 550,000 transistors onto a chip. This chip-based system is five times more powerful than the same company's more conventional AI computers.

Such developments may appear to have little relevance to Third World problems, but it is interesting to note that expert systems have already been applied by aid agencies to the task of logistic planning for famine relief in Ethiopia.

"Expert systems have already been applied by aid agencies to logistic planning for famine relief in Ethiopia."

Meanwhile, the pace of computer development continues to accelerate. With the advent of the "intelligent chip," many analysts believe that the computer of the future will be a hybrid, capable both of number-crunching and symbolic processing. Apollo Computer and Sun Microsystems have unveiled computer workstations that incorporate 32-bit chip processors and that can run numeric and logic processes in parallel.

The promise of AI has been recognized in Japan, where the Ministry of International Trade and Industry, supported by such companies as Fujitsu, NEC, and Hitachi, is undertaking its 10-year Fifth Generation computer program. One objective is to develop "inference machines" that can speed the handling of AI software, as well as database-management machines able to handle enormous quantities of facts. Although a number of other countries are developing their own AI programs, opinion is still divided on whether AI systems will be commercially available soon. If they are, it is still an open question whether they will be confined to niche markets or, as some argue, will begin to reshape the way that mainstream computing is done. However, when Texas Instrument's W. Joe Watson predicted that "AI *per se* will lose its identity within about 5 years," he was reflecting the view, and one which is gaining ground within the IT industry, that artificial intelligence will become an integral part of computing in the 1990s.[40]

No matter how much enthusiasm there may be for AI, however, like all products of human brains, AI systems are far from infallible. Du Pont had to add rules to one of its systems, designed to diagnose problems at a chemical plant, because it failed to spot a clogged valve—caused when a bird fell into an open hatch and drowned. In the environmental field, too, computers analyzing atmospheric data from polar satellites failed to recognize the developing "ozone hole" over Antarctica because they had been programmed to dismiss such a large decrease in ozone as spurious. Intelligent monitoring systems are now being designed to analyze errors, and report on such anomalous conditions.

"Computers analyzing atmospheric data from polar satellites failed to recognize the developing 'ozone hole' over Antarctica because they had been programmed to dismiss such a large decrease in ozone as spurious."

Neural computers

As the emphasis moves from "data into information" towards "data into knowledge," the limits of the present generation of computers may well be reached. But various technologies now on the horizon promise to enormously expand the power and speed of the next generation of computing systems.

Presently, the development of new chips represents the refinement of today's chip architecture, but the "neural" computer heralds a completely new approach. While conventional computers act on information serially, scientists at companies like Hecht-Nielsen Neurocomputer, Revelations Research, and Synaptics now aim to simulate the way neural nets function in the brain and to produce computers able to handle many streams of data simultaneously. The speed and power of early neural-net systems suggests that the goal of producing a computer with true intelligence, capable both of dealing with unforeseen situations and of synthesizing new knowledge from unstructured data inputs with little or no outside help, is achievable.[41]

"So-called 'neural-net' computers 'think' in the same way as the human brain. The speed and power of early neural-net systems suggests that the goal of producing a computer with true intelligence, capable both of dealing with unforeseen situations and of synthesizing new knowledge from unstructured data inputs with little or no outside help, is achievable."

The development of neural processing has accelerated research into optical computers—computers which operate with light rather than electrical currents. The speed of electronic computers is constrained by the time it takes electrons to move from point to point. Further, virtually all electronic computers have to perform functions in series (that is, one computational process after another). However, the optical equivalent to an electrical circuit, an optical fiber, can transport light much faster than electrons can move in chip circuits. Further, if one thinks of a lens as the optical equivalent of a transistor, it has the potential to resolve millions of points, so offering the possibility of making a large number of simultaneous computations.

The "optical switch" is the photonic equivalent of a transistor. AT&T's Bell Laboratories is researching three such devices, which may form the basis of the next generation of computers. The Aerospace Industries Association of America is forming a consortium with the longterm goal of developing an optical computer one thousand times faster than existing mainframes (present Cray supercomputers can perform up to 1,000 billion operations per second).[42]

Meanwhile, however, conventional forms of computing are well entrenched, and it may be that new "superconductor" materials will ultimately transform electrical computing, much as transistors once did. Whereas superconductors once had to be cooled to absolute zero ($-459°F$) to conduct electricity with no resistance, AT&T, IBM, and Toshiba and other companies are finding new superconductors which can operate at temperatures approaching room temperature. Experimental work at Rochester and Cornell Universities indicates that these materials eventually will be able to transmit data 100 times faster than optical fibers.[43]

A clean technology?

Clearly, IT is advancing rapidly on a very broad front. Given that the transfer of earlier technologies to the Third World has led to disasters such as the Bhopal gassings, it is worth asking whether IT is likely to lead—directly or indirectly—to major environmental problems.

> *"While they may be orders of magnitude cleaner per unit of output than most 'sunset' or 'smokestack' industries, IT industries do use chemicals and other hazardous materials."*

The semiconductor and computer industries represent the heartland of the "sunrise" sector of industry. In the public mind, their image is often derived from magazine pictures of the highly sanitized "clean rooms" in which microchips are assembled. State pollution officials in Silicon Valley also tended to think of the industry as "squeaky clean," as "not so much factories, but something like insurance company offices." But, while they may be orders of magnitude cleaner per unit of output than most "sunset" or "smokestack" industries, they do use chemicals and other hazardous materials. To take just one example, by 1979 Silicon Valley was already using 64,000 cubic feet of arsine each year, a toxic gas used as a weapon in the First World War. Silicon Valley has already suffered significant groundwater pollution resulting from the leakage of solvents and other chemicals from underground tanks used by the IT industry.

Recent research has also suggested that there are potential health hazards in the "clean rooms" themselves. Advanced Micro Devices, AT&T, Intel, National Semiconductor and Texas Instruments are just some of the companies that have removed pregnant employees from microchip production areas following concern, originating with Digital Equipment, that exposure to such materials as nitric and sulfuric acids may trigger miscarriages. A five-year study of pregnant women at Digital found that they had experienced a miscarriage rate of 39 percent, compared with a national average of roughly 20 percent.

> *"More than any other industry, the information technology industry seems to have side-stepped the 'Limits to Growth' arguments."*

Environmentalists rightly reason that the ubiquitous chip is largely constructed from silicon, itself derived from silica, an element with which the Earth is particularly well endowed. Furthermore, the dramatic miniaturization of IT components, with the processing power of a major computer now crammed onto a microchip, has meant that ever smaller quantities of raw materials are converted into ever more powerful computing or communication equipment. More than any other industry, the information technology industry seems to have side-stepped the "limits to growth" arguments.

Where the IT industry has run into problems, it has often been able to provide its own solutions. One semiconductor manufacturer in Silicon Valley spent over $50 million on cleanup and evacuation activities associated with repeated leaks of freon gas into its wafer-fabrication area. It is now experimenting with a mass-spectrometer-based atmospheric monitoring system. The heart of the system is a Perkins-Elmer product that started out as a component of instruments the company developed for Trident nuclear submarines. Called ICAMS, it detects parts-per-million concentrations of up to 25 selected compounds at 50 locations, using one analyzer and one IBM PC to manage the data.[44]

A World Resources Institute report, *The Sky is the Limit: Strategies for Protecting the Ozone Layer*, identified the use of chlorofluorocarbons (CFCs), and

CFC-113 in particular, in the manufacture and cleaning of electronic components made with certain plastics, as a further cause for concern. However, increasingly comprehensive regulations in the U.S. controlling the use and disposal of CFCs have created strong incentives for recycling (which can cost up to 12 times more than straightforward replacement) and for developing CFC substitutes.[45]

There is one form of pollution which the IT industry may find difficult to control, however. Some areas of Japan, it has been reported, are now suffering from "electronic smog," caused by electromagnetic waves emitted by equipment like personal computers and game machines. These emissions can trigger uncontrolled responses in other machines, including airport radar systems.[46]

The other side of the coin is that computerization, particularly where sophisticated monitoring and feedback loops are used, has generally improved performance standards in the related fields of health, safety, and environment. It has also increased energy and resource efficiency. The next two chapters look at a number of areas where IT has played a positive role, and more importantly, areas where it potentially could have important contributions to make.

19

III.

SATELLITE REMOTE SENSING

The most striking example of the application of IT in environmental management to date has been satellite remote sensing. Although the United States pioneered in this area with the Landsat program, a number of other countries have now developed significant remote sensing capabilities. This chapter tracks the development of remote sensing as one example of the IT industry's impact on the provision of environmental data. It also focuses on the effects of commercialisation; the provision of services to developing countries; technology transfer; and the dissemination of data.

> *"The most striking example of the application of IT in environmental management has been satellite remote sensing."*

While satellite remote sensing is by no means the only source of environmental data, it can provide the greatest volume of data on a synoptic scale (i.e., global or continental). The original impetus for its development came primarily from military requirements. Even during the Second World War, high-flying aircraft were used to carry out economic reconnaissance (or "econ recon"), with the aim of assessing a country's ability to sustain hostilities. Soon, however, the data available from high-

resolution aerial photography were superceded by information from new sensors exploiting the reflectance of specific wave-lengths of light and near infrared sections of the electromagnetic spectrum.

Although a desire to promote sustainable economic development was not the primary driving force behind the evolution of America's remote sensing capability, military applications have turned the crude art of the 1850s and 1860s, when panchromatic cameras were flown aboard tethered balloons, into a highly sophisticated area of science and technology.[47]

Satellite remote sensing technology has developed very rapidly in recent years. Only three years after the launch of the first (Soviet) artificial earth satellite in 1957, the first (U.S.) meteorological satellite, TIROS-1, was launched, laying the foundation for the Global Observing System of the World Weather Watch. Later satellites were placed in a "geostationary" orbit—where they travel at an altitude of around 23,000 miles at the same speed at which the earth rotates, thus maintaining their position above a given point. Today, the system provides real-time weather information that is broadcast to over 1,000 stations in more than 125 countries. Satellite systems, coordinated by the World Meteorological Organization, now comprise the backbone of the geostationary weather satellite system. Weather

21

sometimes can have an important effect on the economy, especially on the agriculture of a country and weather trends and other climate data almost certainly will have a significant role to play in sustainable development.

In 1968, the National Aeronautics and Space Administration (NASA: see page 67) took the lead in creating the Landsat program, the main focus being on the remote sensing of terrestrial resources. The earth resources technology satellite (ERTS) program was initiated in the 1960s, subsequently changing its name to the Landsat program. Landsat 1 was launched in 1972. Orbiting some 570 miles above the earth, the satellite circled the earth 14 times a day and produced a complete set of images of the whole globe every 18 days.

With two sensors aboard, Landsat 1 proved a considerable success. Plans were developed, under the "open skies" policy of U.S. civilian space projects, to provide data to other governments. No charge was made for the operation of the satellites and many countries began to use the remotely sensed data. Industrialized countries like Canada, Japan and countries in the European Community used the data in a growing range of applications. Countries with developing science and technology capabilities, like Brazil, Argentina, India, Kenya, and Thailand, saw Landsat data as a key to accurate maps and up-to-date natural resources inventories and crop monitoring programs. By the program's tenth year, sales of data had increased 40-fold. Twelve ground receiving stations had been built to provide a distribution network supplying data to over 100 countries.

NASA was responsible for the design and launch of the Landsat satellites and sensors, while the National Oceanic and Atmospheric Administration (NOAA: see page 68) was responsible for their operation and data distribution. With the rapid evolution of IT technologies and sensor capabilities, NASA was able to introduce a new sensor on the Landsat 4 and 5 satellites. The relatively simple Multispectral Scanner (MSS) was complemented with the Thematic Mapper (TM). With an approximate resolution of 90 by 90 feet, compared to the 230 by 230 feet achieved by the MSS, coupled with a greater spectral resolution, the TM images enabled analysts to discriminate, for example, between wheat and barley.

With this extended capability came a euphoria, with the result that the technology was often oversold. Meanwhile, research funding was channeled into commercial areas which would give the best return on value-added services, such as mineral surveys and crop yield estimates. This tended to shift interest away from those areas which were of most interest and value to Third World countries.

"Research funding was channeled into commercial areas which would give the best return on value-added services, such as mineral surveys and crop-yield estimates. This tended to shift interest away from those areas which were of most concern and value to Third World countries."

Happily, however, the U.S. Agency for International Development (USAID) spotted the potential of these new techniques early, starting its own dissemination program in 1971. Even earlier, in 1962, the International Remote Sensing Seminars were begun by the Center for Remote Sensing Information and Analysis, based at the University of Michigan's Willow Run Laboratories (WRL). In 1973, WRL assumed its present name, the Environmental Research Institute of Michigan (ERIM; see page 63). Programs developed by organisations like ERIM and USAID have laid the foundations for training and project work in more than 30 developing countries.

USAID also sponsored programs in specific developing countries. In Thailand, for example, it launched three programs. These programs helped fund Thai scientists to study remote sensing applications. A training program was established, for example, in which 10 Thai scientists were trained in the United States and around 70 in Thailand. And support was also given to build up equipment for receiving and processing remotely sensed data. This helped Thailand begin an inventory of its forests in 1972. It was soon realized that the existing deforestation rate of around 6,650 square kilometers a year was unsustainable, and the Royal Forestry Department began a reforestation program. Landsat data are now used to monitor the restoration and cropping program on a three-yearly basis.[48]

Deforestation, however, is a planned policy in some countries like Brazil and Argentina. In many cases, Landsat data simply enable more efficient stripping of jungles and forests. But, interestingly, Brazil's neighbors have used Landsat data as evidence to illustrate to Brazil the negative impact that its deforestation and industrialization policies are having, both on Brazil itself and, increasingly, on the entire region.

Environmental information in isolation is no guarantee of sound environmental management, however. In 1977, USAID helped the Costa Rican government devise a national forest survey using Landsat data. The survey showed that the current levels of deforestation were unsustainable, indeed that they were—and are—undermining the country's ability to grow sufficient food and cash crops. The Costa Rican government responded rapidly: it moved its remote sensing specialists to other positions.

By the time Landsat 3 was launched, ground receiving stations around the world were processing the data. Their application to geological exploration for oil and minerals, cartography, vegetation mapping, and crop prediction was being heavily researched and—particularly in the commercially rewarding area of geological exploration—increasingly applied. But the oil and natural resource companies took six to eight years to develop the sophisticated proprietary systems needed to turn raw satellite data into information that they could use directly.

One example of the ways in which remote sensing data are moving into the economic mainstream is provided by Cropix, a company started by potato farmers in Oregon that predicts potato production. "Knowledge of crop yields obtained by remote sensing could mean the difference between profit and loss for a farmer," said one user.[49] "For example, potatoes grown along the Columbia Basin in 1983 sold for $80 per ton, and then ninety days later when the harvest was completed, prices soared to $130 per ton due to depressed yields." This sort of information can save individual farmers hundreds of thousands of dollars.

In water-starved Arizona, on the other hand, the State Water Resources Department is using Landsat data and $300,000 worth of image processing equipment to pinpoint "water-rustlers."

Soon, too, the eyes in the sky that scrutinized farmland, areas of desert encroachment, and receding rainforest, were also trained upon the oceans. The feasibility of oceanic remote sensing was tested by the U.S. GOES-3 and Seasat satellites, which were followed by the Nimbus-7 satellite, whose Coastal Zone Color Scanner (CZCS) provided ocean images that proved to be of considerable value to the fisheries industry. Line drawings of ocean-color boundaries were distributed free to albacore tuna fishermen and to researchers about once every ten days through the Scripps Institution of Oceanography at the University of California. Without effective management of such fisheries, satellite data could obviously lead to overexploitation. Similar sensors are planned for the European Space Agency's ERS-1 and Japan's MOS and, later, JERS-1.[50] EOSAT also plans to fly an ocean color instrument, Sea-WiFS, in conjunction with NASA on Landsat 6.

The importance of the polar orbiting weather satellites operated by NOAA as a source of data for large-scale vegetation mapping was also recognized by researchers at NASA's Goddard Space Center.[51] The Advanced Very High Resolution Radiometer (AVHRR) sensor is now inappropriately named, given recent sensor developments offering much higher resolutions. It was designed to monitor global weather patterns with an effective resolution of either 0.7 by 0.7 or 2.5 by 2.5 miles. Data from the AVHRR have been processed at the Goddard Space Flight Center (GSFC) into vegetation index data which are used for vegetation monitoring in Africa. The GSFC vegetation index software provided the basis for the design of a remote sensing system to support food security monitoring and desert locust swarm forecasting in Africa. Other studies using these data have evaluated deforestation trends in Brazil and vegetation cover trends in the Nile Delta.

One long-running international environmental data collection progam based on two NOAA satellites is ARGOS (see page 56). This system has been continuously operational since 1979 and involves cooperation between NASA, NOAA, and CNES, the French Space Agency. The data are used for a wide range of activities, including the monitoring of a network of oceanographic buoys and the tracking of wild animals on land or at sea.

There are two immediate reasons why the NOAA satellite AVHRR remote sensing data may be preferred to those from earth resources satellites for applications requiring frequent coverage of large areas. First, this sensor has a coarser ground resolution, providing a cheap, but effective synoptic view, and the NOAA system can monitor a large area on a frequent basis. Second, the weather satellites are part of an operational system. The importance of this fact was brought home to anyone dependent upon Landsat data as the working life of Landsat 5 began to run out and Landsat 6, with an exciting array of new, more powerful sensors, was grounded by a combination of financial pressures and the moratorium on Shuttle missions in the wake of the *Challenger* disaster.

Meanwhile, Satellite Hydrology Inc. (see page 70), one of the many value-added consultancies now helping to develop the market for satellite imagery, processes data supplied directly by NOAA as part of the LEAP (Landsat Emergency Access and Products) Program. LEAP was designed by NOAA to provide instant access to remotely sensed data for such agencies as the National Guard and the Army Corps of Engineers, which need to respond to natural disasters and civilian emergencies.

NOAA data are also used in the U.N. Food and Agriculture Organisation (FAO) African locust early warning program. However, although NOAA data give a regular synoptic view of those areas where locust breeding occurs, scientists have found it difficult to distinguish between potential breeding areas and shadow. High resolution pictures, such as those produced by SPOT, the French commercial satellite, would help, but are more expensive. The locust control program, for this and other logistical and political reasons, has made very little impact on the African locust problem.[52]

In 1982, the FAO presented *Remote Sensing Applied to Renewable Resources* to the Second U.N. Conference on the Exploration and Peaceful Uses of Outer Space, held in Vienna.[53] This report described the potential of remote sensing for inventorying natural resources, monitoring change in natural systems, and providing early warning of natural disasters. This drew on earlier work, including a report in 1981 on tropical forest assessment in Africa.[54] In 1986, another report (*System Definition of Africa Real-Time Environmental Monitoring Using Imaging Satellites—Project ARTEMIS*[55]) described the first operational international project for monitoring renewable natural resources. The ARTEMIS system is designed for the regional monitoring of vegetation and precipitation.

On a larger scale, the Global Environment Monitoring System (GEMS) was set up in 1974 by the United Nations Environment Programme (UNEP) to provide a central mechanism for monitoring climate, the long-range transport of pollutants, renewable resources and the oceans and other environmental parameters on a global basis. Incorporated in GEMS is the Global Resource Information Database (GRID) system, designed to coordinate the enormous input of environmental and natural resource data from GEMS and other data bases. With applications software developed by ESRI, information held in a number of databases is integrated into a geographical information system (GIS), which draws heavily on data from the NOAA and Landsat satellites. The primary advantage of the system is that satellite data can be integrated with almost any other geographically based dataset. Ultimately, according to Michael Gwynne, Director of GEMS in Nairobi, the goal is to have a GRID capability in every country. The falling cost of computers and other information technologies is seen as an important facilitating trend.[56]

The Government of Senegal, with funding from USAID and technical assistance from the Remote Sensing Institute of the South Dakota State University, set up a pilot GIS to assess its performance for a national plan for the management and optimal utilization of Senegal's natural and human resources. This GIS, which covers the economically important groundnut basin, holds information about soils, climate, land-use, forest and pasture reserves, crop use intensity, crop management data, and more. The GIS can be interrogated and interpreted on questions such as erosion control, crop suitability, and population distribution.[57]

After work with Landsat images, the Indian Space Research Organisation (ISRO) has developed its own plans to further develop its National Natural Resource Information System. An Indian satellite will provide data to the system, which will be processed to provide information on a wide range of management issues, from agriculture and forestry to marine resources.[58]

In highly industrialized or agriculturally intensive countries, such as those of Europe and North America, advanced GISs which incorporate remotely sensed data have been developed, and these are often highly integrated with the planning process. For example, as the OECD has reported, Norway and France have both developed systems of resource accounts which internalise the value of natural resources and provide budgets for their use and development.[59]

The synoptic view offered by satellite data makes them a driving force behind some of the major emerging global research programs operated by such agencies as NASA and NOAA. For example, satellites were used extensively during NASA's Amazon Boundary Layer Experiment (ABLE), undertaken as part of the Global Tropospheric Experiment (GTE), in conjunction with data from aerial and ground surveys. The objective was to assess the role of the Amazonian forests in global geochemical cycles.

When the Research Briefing Panel on Remote Sensing of the Earth reported, it noted that "until the flight of satellites, there were no techniques for long-term, global, synoptic measurements of processes in the atmosphere, oceans, and solid earth. Now we are on the verge of establishing a system of remote sensing instruments and earth-based calibration and validation programs that could provide such a data set. With the concurrent development of numerical models that can run on supercomputers, we have the potential of achieving significant advances in understanding the state of the earth, its changes, feedbacks, interactions and global trends on time scales ranging from days to centuries."[60] A key question, however, is whether the necessary investment will be made to establish such a system, and make its outputs available to those who need them most?

In 1984, NASA produced a two-volume working report describing an Earth Observing System (EOS). Based on a seven-sensor system able to conduct both passive and active experiments, the intention is that EOS would be flown on a polar platform sometime in the 1990s. The expected data output rates would be prodigious, in the range of 700-800 Megabits per second (Mb/s), way in excess of the current 300 Mb/s capabilities of the Tracking and Data Relay Satellite System (TDRSS), which tracks such operating orbiting satellites as Landsat and relays data to ground-receiving facilities.

The evidence suggests that future developments in this area are likely to continue to be driven by the military sector, whose activities are not usually seen as contributing to environmental goals or sustainable development. As *High Technology* has pointed out, the Strategic Defense Initiative (SDI) is already producing advances that could help unlock non-military applications. The Lasercom system, developed by the Massachusetts Institute of Technology's Lincoln Laboratory as part of the SDI program, can support data exchange at the rate of 225 Mb/s. This will have to increase to 1 Gigabits per second if the SDI program is to meet its objectives, coincidentally the very rate of data transmission that NASA forecasts will be needed for low-orbit earth resources satellites by the year 2000.[61] Clearly, these potential benefits do not represent an environmental case for SDI, but if the SDI program goes ahead they should not be ignored. Neither should the huge volume of data collected solely for military or other intelligence purposes.

Programs like the Earth Observing System, as John Estes and Jeffrey Star of the University of California at Santa Barbara (see page 70) have noted, "can produce an evolutionary improvement in our understanding of our planet," promoting multidisciplinary research on an international scale.[62] As such research programs evolve, so the volume of data that will need to be handled will grow almost exponentially.

"Environmental researchers are going to have to 'drink from the firehose' of satellite data streams."

Environmental researchers, as the Research Briefing Panel on Remote Sensing of the Earth put it, are going to have to "drink from the firehose" of satellite data streams. Scientists have looked at only 10 percent of the data collected by satellite, and have closely analyzed no more than one percent.[63] NASA, not surprisingly, sees EOS not so much as a remote sensing program as an information sciences project, recognizing that major advances in information sciences and technology are a necessary condition of major progress in global environmental monitoring.

Various organizations will deal with such problems and resources in different ways. The World Data Centers established by the International Council of Scientific Unions (ICSU), for example, take a relatively centralized approach to the international exchange of data. Other programs, including UNEP's GEMS and INFOTERRA programs, tend to be sectoral or problem-oriented. And a third category of databases, including GRID, are likely to be much more fully distributed. The key to advances in data management, however, as NOAA's NOAAPORT and NOAANET projects confirm, will be the user-friendliness of the end-products. Clearly, IT will have an increasing role to play in the distribution of environmental data to end-users, and the integration of this information into environmental management processes.

Enormous quantities of environmental data have been collected in recent years. NOAA gathers worldwide environmental data about the oceans, earth, air, space, and sun and their interactions, with a view to describing and predicting the state of the physical environment. Once used by NOAA or some other collecting agency, environmental information is passed to four service centers maintained by NOAA, where it is made available though the National Environmental Satellite, Data, and Information Service (NESDIS: page 67).

New systems are being devised to exploit these data. ERDAS, to take one example, has worked closely with government agencies and the private sector to develop image-based techniques for environmental applications that can be run on micro computers. Databases have been constructed and installed for the U.S. Forest Service, U.S. Geological Survey, the United Nations, and various research institutes. In conjunction with ESRI, ERDAS has conducted technology transfer programs for UNEP in Africa. In a collaborative effort, ERDAS and ESRI developed a link for data transfer between micro computers running ERDAS and mini computer-based ARC/INFO systems for the GEMS GIS—GRID.

Inevitably, the rate of technological advance has posed its own problems for developing countries. The new TM sensor, fitted to the Landsat 4 and 5 satellites, transmitted its data at freqencies different from those used by the MSS. Ground stations built at a cost of around $10 million to receive MSS data are not able to receive TM data. As a concession, NASA fitted both the TM and MSS sensors to Landsats 4 and 5. Unfortunately, Landsat 4 developed a power problem soon after launch, and the MSS was switched off. When Landsat 3 stopped operating in March 1983, non-TM ground stations had to wait until Landsat 5 was launched a year later before data came on stream again. Few developing countries were affected by this as many restrict their use to cheap archive data for non "real-time" applications such as mapping.

This hiatus provided a perfect opportunity for France's *Systeme Probatoire d'Observation de la Terre* (SPOT: see page 71) to stake a claim in what some analysts believe will be a $2-$4 billion market by 2000. The SPOT satellite was launched into its 520-mile high orbit aboard an Ariane-1 rocket early in 1986, the first operational commercial earth resources satellite. Gerard Brachet, as Director-General of SPOT Image, has forecast that SPOT's market share could rise slowly to account for a modest $100 million of the market forecast for the turn of the century. Initial sales of SPOT data were lower than expected, although they are on a rising trend.

With a ground resolution of 30 by 30 feet for panchromatic ("black and white"), the military interest in SPOT imagery could be considerable, particularly from Third World countries either unable to afford their own satellite or unable to extract suitable images from a friendly superpower. As far as natural resource management goes, however, SPOT expects the main demand for agricultural data to come from industrialized countries, where the patterns of agriculture are well suited to the detailed images SPOT can provide. A secondary market is also expected to develop in fisheries management.

But there is a major disadvantage to Third World countries in the technological advances that the SPOT satellite has incorporated in its design. To reduce power demand on the satellite, the SPOT sensors can be switched off when not required. Consequently, the resulting remote sensing coverage is selective, generally covering areas where clients are prepared to pay. The Landsat's indiscriminate sensing method serves developing countries better, in that data are generated over these countries regardless of whether there is immediate demand.

The emergence of the SPOT program has clearly come at a difficult time for the Earth Observation

Satellite Company (EOSAT: see page 62). Created as a joint venture between the Hughes Aircraft Company and the RCA Corporation, EOSAT was designed to take the Landsat program into the private sector.

The Landsat Remote Sensing Commercialization Act of 1984 provided for the operation of Landsats 4 and 5 with funds from NOAA's 1986/87 budget, and a further sum of $295 million was earmarked for the development and launch of Landsats 6 and 7. In 1985, EOSAT's sales amounted to $20 million, but the cost of Landsat data was skyrocketing. Indeed, Warren R. Philipson of the Cornell Laboratory for Environmental Applications of Remote Sensing (CLEARS: see page 59) has pointed out that the cost of a single Landsat TM scene in digital form exceeds the basic cost of a microcomputer.[64] Further, for most applications, the data require further processing and the cost tends to offset any savings from the multiple use of the data in more than one application. As a result, the data are well beyond the reach of most potential Third World users.

"Remote-sensed data are well beyond the financial reach of most potential Third World users."

EOSAT's ability to compete in commercial remote sensing has been severely hampered by the delayed launches of Landsats 6 and 7, problems that have been aggravated by the rescheduling of government funding for the development and launch of these two satellites. If Landsat 6 does finally get into orbit in 1989 or 1990, EOSAT will still have a difficult task persuading customers that its operational system can support extensive, ongoing monitoring programs.

Efforts are being made to stimulate the value-added industry as the real marketers of satellite data. EOSAT's 1986 directory of Landsat-related products and services listed over a hundred companies operating in this field, and the hope is that they will extend the range of applications and spread the cost of the data among a growing array of end-users.[65]

The value-added sector is pulling in some well-established companies, including Kodak (see page 66). Kodak was an unsuccessful bidder for the EOSAT operation, but has since decided to move into the image processing and remote sensing field, and has set up a subsidary, KRS, to do so. Already well established in the optical disk market, Kodak may be one of the organizations to once again lower the entry costs to image processing significantly.

"Notwithstanding the falling costs of processing equipment, the price of remotely sensed data is increasing."

Notwithstanding the falling costs of processing equipment, the price of remotely sensed data is increasing. By the time Landsat 5 was launched, the U.S. had had over ten years of experience of operating its "open skies" remote sensing program. Sales of data had grown from around $230,000 in 1973 to about $2.9 million in 1983. However, it cost $40 million to operate Landsat in 1984, and the Landsat 4 satellite cost $573 million to develop and build and a further $36 million to launch.

It is not therefore surprising that the cost of data has increased with demand. For the first ten years, the US government charged an annual "licence fee" of $200,000 annually for each satellite data ground receiving station. In 1983 this was increased to $600,000. Prior to 1981, a single multispectral picture cost about $200. By 1985 the cost had risen above $700. TM data cost almost $3,000 per scene when it became available in 1983, but rose to $4,400 by 1985.

Although the demand for satellite data is growing, so are the number of competitors for the market in remotely sensed data. For example, in 1987, Japan launched its own earth resources satellite, the MOS-1, which will meet many of its own needs, and possibly sell data to other countries. The European Space Agency has plans to launch its satellite as have countries like Brazil, China, India, and the Netherlands. The USSR already has remote sensing capabilities, but sells little data commercially. The Netherlands has plans for a satellite in co-operation with Indonesia to cover the special problems of the equatorial and tropical belts. Indonesia already uses Japan's GMS weather satellite data to plan cloud seeding operations to support rice crops in the dry seasons.

Clearly, as more operational remote sensing systems look for ways to fund themselves, prices of data will rise and shift the emphasis of its application away from environmental management and towards commercial applications.

Not only does cost inhibit distribution of data, but in the United States there are regulations which give the government the power to restrict the sale of any satellite data which may pose a security risk. Some Pentagon officials already feel that the 5 meter resolution SPOT images fall within this category. Currently, commercial satellites like SPOT and Landsat operate under the "open skies, open access" policy—one which gives any country the right to collect data over any area, and distribute that data to any user. This policy has not met with the general approval from developing countries, many of whom fear that they may be excluded from sources of information about their own countries.[66]

One interesting proposal advanced by Joan C. Hock and Jennifer Clapp of NOAA, and John H. McElroy of Hughes Communications, Inc., is to establish a new private international entity for earth observations, dubbed ENVIROSAT, along the lines of INMARSAT or INTELSAT (see page 58).[67] The core public data set, it is suggested, would be distributed free to anyone needing it, whether or not they contribute to system costs, as is the case for current weather satellite products. These data would be treated as "loss leaders," pulling in users for premium, value-added services. For-profit subsidiaries could participate in value-added markets.

Even with an ENVIROSAT, there would be a continuing need for national programs, much like India's and others planned in countries like Brazil. However, although entry costs may be falling with cheaper equipment for such large scale installations, hidden costs in training, servicing, and maintenance may continue to represent a significant barrier to the development of remote sensing capabilities in developing countries.

"Although entry costs may be falling with cheaper equipment for such large scale installations, hidden costs in training, servicing, and maintenance may continue to represent a significant barrier to the development of remote sensing capabilities in developing countries."

The availability and quality of the professional and technical infrastructure needed for such technologies are critically important. As T. Woldai of the Netherlands-based International Institute for Aerospace Survey and Earth Sciences (ITC) commented on an earlier draft of the present report: "In my recent visits to China, Indonesia, and Thailand, (I found that) one of the main problems faced by remote sensing users or computer scientists is the maintenance of computers or data processing facilities. We may talk about advances in software development and hardware gadgets, but there is a major crisis because of the lack of better trained people in this field. Many developing countries cannot afford to spend many thousands or millions of dollars unless they are sure they have someone at home to see that their computer facility is properly maintained." As Chapter IV shows, these concerns are not restricted to remote sensing.

IV.

SOME OTHER APPLICATIONS

Many Third World governments now recognize that IT can play a key enabling role in development. Telephones and telex facilities, explained Dr N.M. Shamuyarira, Zimbabwe's information minister, "are essential parts of that human organization required to tackle mass poverty, illiteracy, disease and general development in Africa." Indeed, a recent study carried out by the International Telecommunications Union (ITU) in the Philippines indicated that the benefits of investment in telecommunications outweighed the costs by more than 30 to 1 in the health sector, and by more than 40 to 1 in the agricultural sector.[68]

As a result, there is growing interest in the Third World potential of IT. One of the pioneering organizations in this field has been the United Nations Industrial Development Organization (UNIDO). When UNIDO reviewed the microelectronics industry in the Third World, it focused predominantly on countries in South America and Asia, including Venezuela, South Korea, India, Pakistan and Bangladesh.[69] Not surprisingly, it found dramatic differences among them.

South Korea, a net exporter of electronic products, was the only country covered whose output of such products was significant in world terms. India has also been increasingly active in consumer and industrial products and components, but was not self-sufficient in any of these areas. Although Venezuela is virtually self-sufficient in telephone production, other sectors based on microelectronics are dominated by foreign based companies, including IBM, Philips, Siemens and Hewlett-Packard. Pakistan has been placing a growing emphasis on investment in microelectronics, with an emphasis on products for export. Bangladesh, on the other hand, is still at the assembly stage in consumer products, using imported products. Only a handful of Bangladeshi companies are using integrated circuits, and most of these are dependent on technical collaboration agreements with foreign companies.

To date, most non-military IT applications in the Third World have been clustered in such areas as computing and data processing, particularly in the public sector; in telecommunications and broadcasting; in process control; and in other public services, including energy supply, transportation, meteorology and medical services.

It is impossible to say whether, on balance, IT has moved Third World economies towards—or away from—sustainable patterns of development. But, since sustainable development is not yet a central policy objective in such countries, the assumption must be that, while IT has helped drive forward a broad range of development activities, relatively few will have been designed for environmental sustainability.

29

Overall, although the phrase "sustainable development" enjoys increasing currency, the concept remains just that, with relatively few real-world cases to point to. In these circumstances it is very difficult to find well established, "operational" applications of IT in the sustainable development field. Nonetheless, as Chapter II reported, mainstream IT capabilities are increasing dramatically in the industrialized world, while the entry costs for IT users—wherever they are located—are falling rapidly. These trends are opening up a broad range of potential applications, not least in environmental management.

> *"In these circumstances it is very difficult to find well established, 'operational' applications of IT in the sustainable development field."*

Taking a longer term perspective, it is clear that remote sensing and other forms of IT offer important new tools for defining and addressing some of the key challenges highlighted by the World Commission on Environment and Development (WCED). *Our Common Future* identified a number of "common challenges" facing humankind, including the management of: (1) population growth; (2) food security; (3) species and ecosystems; (4) energy; (5) industry; (6) urbanization; and (7) the global commons. To demonstrate the potential of IT in sustainable development, we now consider a small selection of existing and potential applications under each of these headings.

The world's human *population* breached the 5 billion mark in 1987, with current forecasts—generally prepared using computer models—suggesting that it will reach 6.1 billion by the year 2000 and perhaps 8.2 billion by 2025. This growth in numbers is unquestionably the most serious non-military threat to the prospects for sustainable development.

High population growth rates, as the World Commission summarized the problem, "already compromise many governments' abilities to provide education, health care, and food security for people, much less their ability to raise living standards. This gap between numbers and resources is all the more compelling because so much of the population growth is concentrated in low-income countries, ecologically disadvantaged regions, and poor households."

Remote sensing data have been used in population census work in a number of Third World countries, including poorer nations such as the Sudan and wealthier ones like Saudi Arabia. Given the growing importance of population:resource linkages, it is interesting to note that the objective of the Sudanese work was to analyze the links between population densities and the process of desert formation, or desertification.[70]

Family planning services are not universally acceptable, but the evidence suggests that where they are available demand can grow rapidly. The problem in many developing countries is that these services are operated in isolation from other fertility-related programs, including those focusing on nutrition, public health and preschool education. IT can help bridge this gap, however. A health and nutrition program in Nepal run by the Save the Children Fund developed database management software to maintain records for several thousand children. Although computerization is not necessarily the answer to Third World health care problems, computerised management systems can help ensure that family planning services are provided to families alongside other services.

In a country like Thailand, where fewer than 5 percent of villages have doctors or trained nurses, the existing network of government health centers is inconvenient for the majority of the population. Television broadcasts of programs like the "Happy Vasectomy Show" have helped get the message that "Many children make you poor" across to a hitherto remote but important audience.[71] Such programs have helped cut Thailand's annual population growth rate from around 3.2 percent in the 1960s to about 1.6 percent today. Many other countries, such as Egypt, Kenya, and Mexico also use television in health and family planning education.

Many IT applications, however, have so far only operated at the pilot scale. But the results achieved suggest what might be achieved given more resources. Often considered a model of the effective application of IT, India's Satellite Instructional Television Experiment (SITE) program was conducted in 1976. It used NASA's ATS-6 communications satellite and a network of 2,400 ground stations in far-flung areas of the country. The educational programs covered such areas as agricultural development, family planning and primary healthcare. However, it should be noted that the project was discontinued after foreign assistance dried up. A similar, but still operational communications program is called PEACESAT. It also uses a NASA satellite and is being used to promote the introduction of new crops and agricultural techniques in the Pacific Basin.

Of course, none of these programs guarantees that subsequent development patterns will be sustainable, but the integration of different sets of data frequently means that development is better planned and coordinated than it otherwise might have been.

And what about IT's ability to help us "think the unthinkable"? It is interesting to see how the global modeling work of the 1970s has spun off a number of computing initiatives with real-world applications. SAI, for example, profiled on page 71, offers a range of world modeling services for government and agencies and business. The company's involvement in the Club of Rome's work helped develop early decision support systems, which its ARISTOTLE/FORESIGHT software system is now opening up for government agency and business use in both developed and developing countries.

At a totally different level, computers are also now being used in many educational and training applications. What, we may ask, would it be like to live in a world which was unsustainable? Although playing with a computer cannot reproduce the full horror of seeing your crops, animals and children die as the desert creeps ever closer, educational software now being developed can give young people a real sense of how different factors combine to produce environmental degradation. "Sand Harvest," a computer game produced by the Centre for World Development Education, is a case in point. It enables young people to get involved in environmental issues through role playing. The focus is on desertification in Mali and they can pick one of three different roles: Nomad, Government Officer or Villager. Annual reports are provided throughout the simulation, to give those involved a sense of the impact that their decisions—and the decisions of the other groups—are having on the environment and on the competing communities who depend on it for their living.

Clearly, *food security* must be a central objective in sustainable development programs. "The application of the concept of sustainable development to the effort to ensure food security," the World Commission concluded, "requires systematic attention to the renewal of natural resources. It requires a holistic approach focused on ecosystems at national, regional and global levels, with coordinated land use and careful planning of water usage and forest exploitation."

The synoptic view provided by satellite remote sensing can bring many benefits in this area, although the successful commercial applications are largely found in the industrialized countries. The Cropix example, described in Chapter III, illustrates what can be achieved. But few of the value-added companies which are moving into the renewable resources market would see attractive commercial opportunities in the Third World. IT, however, *is* increasingly used in Third World environmental management programs.

In Kenya, for example, the USAID-assisted Remote Sensing Department of the Regional Centre in Surveying, Mapping, and Remote Sensing has been using Landsat data, often data which are eight to ten years old, to map areas of East and Southern Africa that have not previously been accurately mapped. These images and maps are being used to assess food production, and were of particular value during the distribution of emergency food relief in 1985. Such maps can be produced from satellite data in three weeks, according to project advisor Allan Falconer, whereas a conventional survey might have taken three years or more.[72] Data provided by the Regional Centre have also helped the Kenya Rangeland Ecological Monitoring Unit (KREMU) to prepare soil erosion hazard and management plans.[73]

IBM has been a partner in work carried out since 1978 by the French Institute of Agronomic Research with the long-term objective of evaluating the biomass content of the soil at the end of the rainy season in such countries as Mali, Mauritania, Senegal, Burkina Faso, Chad, Niger and the Cape Verde Islands. IBM has supplied the software needed to interpret the digital information supplied by remote sensing satellites. The resulting information, and that supplied by the IBM Scientific Center in Cairo, is used to help farmers improve their soil management practices.

31

In Mexico, too, satellite data have been used to delineate the country's potential arable soil resources. The study found 6.3 million hectares in a state of advanced erosion, but gave the Mexican government a more robust basis for future natural resource management decisions.[74]

ARTEMIS is one operational crop and vegetation monitoring system, mentioned in Chapter III, based on remotely sensed data from NOAA's AVHRR satellite. The Assessment and Information Service Center (AISC) of NESDIS has also used AVHRR data combined with meteorological data from the World Meteorological Organization to provide a climatic impact assessment service. Data from disparate sources are combined in a GIS and have allowed AISC to notify countries like Mauritania, Senegal, Mali, Burkina Faso and Chad of impending food shortages, up to one month before crop losses occurred. However, in some cases the technological ability to monitor changes in environmental conditions has outstripped the ability to communicate and act on that information.

> *"In some cases the technological ability to monitor changes in environmental conditions has outstripped the ability to communicate and act on that information."*

Advanced computer techniques like artificial intelligence and expert systems are beginning to find applications in natural resource management. The extremely complex logistical problems associated with the supply and distribution of food aid prompted the development of an expert system which relief organizations can use to plan their programs. More sophisticated systems are being used to manage the production of lucrative commercial crops in the industrialized countries. The COMAX system, for example, helps cotton growers schedule irrigation, fertilizer applications and harvesting. Although the computer hardware and weather station equipment can cost between $10,000 and $13,000, the increased yields achieved can sometimes be repaid many times over in the first year of operation alone.[75]

Although these applications may seem somewhat removed from the immediate resource management

needs in developing countries, they are spurring the development of cheaper advanced GIS and expert system technologies which should find wider applications. The use of the new 32-bit computers to run such software will also make such systems much more "user-friendly."

As far as the conservation of *species and ecosystems* is concerned, a major user of IT is the Conservation Monitoring Centre (CMC), established by the International Union for the Conservation of Nature and Natural Resources (IUCN) in collaboration with UNEP, the World Bank, the World Wildlife Fund and other organizations. CMC's aim is to ensure that development projects are designed with environmental and species conservation in mind. It integrates four monitoring activities, covering the status of animal species, plants, the wildlife trade and protected areas.

CMC makes extensive use of computers. One database, which is being assembled at Kew, England, holds information on some 5,000 plants, 500 of which act as "green glue," covering dry ground and binding erosion-prone soils. Also included are African plants that can be used as natural insecticides, and some that are unpalatable to animals which might otherwise cause overgrazing. Under the Convention in Trade of Endangered Species (CITES), too, every ivory producing country has an annual ivory quota, and each tusk receives a unique reference number. It is CMC's task to keep track of all ivory imports and exports within the CITES system.[76] Again, computers play a key role. CMC is also now working on ways of integrating the information collected by its various component bodies into a GIS format for use in planning and management.

Remote sensing data have also proved invaluable in practical conservation projects. Landsat imagery has been used in support of the White Oryx Project in Oman, for example, which is attempting the reintroduction of the White Arabian Oryx in the Jiddat al Harasis region. Among other computer applications in the area of animal conservation, the International Species Inventory System (ISIS), based in Minneapolis, and NOAH (National Online Animal History) and ARKS (Animal Records Keeping), based in England, help ensure that breeding programmes avoid inbreeding and maximize genetic diversity in captive animals.

IT has found a profusion of applications in the *energy* sector—and will find many more. "A safe and sustainable energy pathway is crucial to sustainable development," the World Commission concluded, but noted that "we have not yet found it." The scale of the challenge is suggested by the fact that to bring energy use in the developing countries up to the levels forecast for the industrialized countries in the year 2025 would imply increasing total global energy use by a factor of five. There are two clear imperatives. More will have to be done with less energy in the industrialized nations, while more energy sources will need to be found to support the transition to more sustainable patterns of development in the developing countries.

One of the new approaches to energy management which emerged in the wake of the energy crises of the 1970s was the Integrated National Energy Planning (INEP) system, developed by Mohan Munasinghe. This energy model operates at three levels: macroeconomic, energy sector, and energy subsector. With the advent of the 16-bit computer, INEP became a powerful, but cheap and flexible tool for energy policy-makers in the Third World. Applied to an analysis of energy supply and demand in Sri Lanka, the model provided some interesting insights. It was used to assess the ability of non-conventional energy sources, such as solar, wind, biomass and minihydro schemes, to reduce dependence on imported oil and local fuelwood. Unhappily, the model suggested that maximum output from the renewable energy sector would meet only around 2 percent of the 1995 electricity generation requirement.[77] Even so, the application of computer modelling allowed the Sri Lankan planners to analyse their country's needs in a comprehensive way. Further, the modelling exercise left the Sri Lankans with a great deal of indigenous expertise to continue the use of computers in energy planning.

The renewable energy contribution is likely to be much higher in some other Third World countries, but it does look as if IT will be more widely used in the fossil fuel and nuclear energy sectors for the foreseeable future. The oil exploration industry, for example, is one of the single largest users of commercial remote sensing data, accounting for about 30 percent of data sales.

With most energy supply options now constrained to some extent by environmental factors, from acid rain to the build-up of carbon dioxide in the atmosphere, computer models are increasingly used to explore the environmental implications of particular energy scenarios. The performance of one such model was reviewed in an earlier WRI report, *A Matter of Degrees: The Potential for Controlling the Greenhouse Effect.*[78] The conclusion drawn was that unless policies are implemented soon to limit emissions of such "greenhouse" gases as carbon dioxide, chlorofluorocarbons and nitrous oxide, intolerable levels of global warming will result. Such models as the WRI Model of Warming Commitment—and more disaggregated models focusing on country-specific or regional policy options for limiting the "greenhouse effect"—will play an important role in identifying workable pathways to the sort of high growth, low energy future called for in both industrialized and industrializing countries by the World Commission.

Whichever development pathways are adopted, energy efficiency can only become a more pressing priority. Groups like the Intermediate Technology Development Group (ITDG) are working on a range of applications of IT designed to bring new forms of energy to Third World users, and to enable them to use energy more efficiently.[79]

National electricity grids, for example, often fail to reach remote areas in the Third World. Without power supplies, they have little opportunity to develop small, local industries which could help improve the quality of life. Hilly and mountainous areas may have access to abundant hydropower potential, but tapping it in such a way that power is produced consistently and reliably can be a problem. The use of mechanical "governors" to regulate the flow of water into minihydro plants often proves unreliable, so ITDG has developed a simple electronic controller which contributes to savings of up to 50 percent on the total capital and running costs for such plants. Because it contains no moving parts, the electronic load controller has proved extremely reliable.

ITDG has also used computers to design fuel-efficient stoves for use in Africa. With flue gas sensors connected to microcomputers running expert system software, ITDG technologists have been able to adjust a large number of design parameters. As a result, they have produced cost-competitive stoves that cut fuel consumption by up to 50 percent. Such

33

stoves have a vital role to play in the many areas of the Third World where deforestation has made fuelwood difficult to find.

As far as *industry* is concerned, most of the relevant IT applications to date have been pioneered in the industrialized countries. New online databases, offered by online services like Dialog (page 61) and EIC/Intelligence (page 62), have been created to alert industry to the availability of "clean" or low waste, low pollution technologies. The European Community, for example, has developed NETT, the European Network for Environmental Technology Transfer. A second initiative currently under development is Polmark, an advanced database and information service on pollution control and waste management technology and markets.

DEC, in conjunction with NTIS (page 69), has produced numerous optical disk databases covering such areas as environmental health and safety, energy, and natural resources. A yearly subscription to these databases costs around $1,200, with data updates (because current CD disks cannot be updated directly) sent out on replacement disks.

The United Nations Environment Programme's Industry and Environment Office, based in Paris, France, has also developed a computerized clean technology database targeted at selected industries in the developing countries, although the cost of creating and updating the database has proved somewhat disproportionate to the benefits achieved to date. A more successful database is that operated by the International Registry of Potentially Toxic Chemicals (IRPTC), another branch of the United Nations Environment Programme. Located in Geneva, Switzerland, this provides users worldwide with computerized information on hundreds of chemicals, including pesticides. The database can be accessed in over one hundred countries in Africa, Asia, and Latin America through a network of "national correspondents." Further, the Natural Resources Development Corporation of Canada is developing a version of the database for use on microcomputers.

To show what is possible at the company level, a computer system in Monsanto's corporate purchasing department lists surplus materials at plant locations. Before any materials are purchased for a plant, the purchasing department checks to determine if they may be available at some other plant. Before this system was implemented, some surplus materials were being disposed as waste. Monsanto also uses a computerised translation system to produce product data sheets, showing a given product's environmental and toxicity performance, in the languages used in key markets.

IT is also making major inroads in the analysis and management of industrial hazards. The cost of analytical equipment has fallen rapidly in recent years, not least because of the way in which the cost of microelectronic components has dropped. As recently as the early 1970s, a typical instrument consisting of a gas chromatograph and mass spectrometer cost almost $1 million and occupied two rooms. Today it sits on a benchtop and sells for about $100,000. That may still place it out of reach of many potential users in the Third World, but the trend is clearly in the right direction.

In the wake of the disasters at Flixborough (England), Seveso (Italy), and Bhopal (India), there has been growing interest in environmental monitoring technologies.[80] The use of field-effect transistors (or Chem FETS) as sensors has been one lively area for R&D. When exposed to a particular pollutant or hazardous chemical, the electronic characteristics of the semiconductor component of the Chem FET change, stimulating a stream of electrons that can be used to activate a recording device or alarm.

Now that the computer hardware exists, environmental software represents another growth area. Technica International, for example, have developed a software package called WHAZAN (World Bank Hazard Analysis Software) which encapsulates the approach outlined in *Techniques for Assessing Industrial Hazards,* a manual they produced for the Bank. The aim was to produce a low cost package (WHAZAN costs around $1,000) to run on IBM-compatible microcomputers. The package includes a library of hazardous materials, and can model the dispersion of liquids and gases, and the resulting human health impact. In fact, so many computer-based expert systems are now available in this area that Battelle have produced an expert system—SOPHIE (Selection Of Procedures for Hazard Identification and Evaluation)—whose sole purpose is to identify appropriate hazard analysis methods for specific plant and safety applications.

The expensive business of environmental control has also generated important new IT applications. Atlantic Richfield (ARCO) has used Apple Macintosh desktop publishing systems to prepare such documents as environmental impact assessments and toxicology reports related to their oil developments. The impact on cost-effectiveness has been considerable. Some thick reports that used to cost around $60,000 to prepare now cost around $13,000.[81]

Computer-aided design (CAD) is an area where IT could also provide new tools for sustainable development. Using CAD techniques, process or product designers can test the performance of their design options on the computer screen, long before metal is first bent or concrete poured. Subsequently, computer-aided engineering (CAE) and manufacturing (CAM) methods help to ensure high standards of quality control in the production or implementation stages. ITDG's stoves, in short, are only the beginning.

As the cost of microprocessors tumbles, so new product generations become increasingly "smart" or "intelligent." Pollution control requirements helped spur the introduction of microprocessors in auto engines, a trend initially resisted by the auto industry—until it recognized the performance and other benefits that could follow. New smart buildings automatically switch the lights on when a person walks into a room, triggered by infrared sensors that detect the slightest movement, and then turn them off soon after the room has been vacated. Given that many Third World capital cities consume more energy than the relevant countries' villages, the potential for improving the energy efficiency of high rise buildings and hotels is likely to be considerable.[82] It is worth pointing out, however, that while computer models now exist which can carry out energy audits on buildings, many other factors feed into the equation which determines the extent to which architects, builders, owners, landlords or tenants pursue energy efficiency goals.

Sometimes, too, IT can have quite the opposite effect to that intended by policy-makers. Consider *urbanization.* The world's economic system, the World Commission noted, "is increasingly an urban one, with overlapping networks of communications, production and trade. This system, with its flows of information, energy, capital, commerce, and people, provides the backbone for national development." But the sheer pace of urbanization in many countries has been the cause of considerable concern. Rural electrification has been proposed as one of several improvements necessary to slow migration by making village life more attractive. Unfortunately, electrification programs often expose communities to the mass media, particularly television, reinforcing the "urban pull."

"The role of advanced telecommunication systems and other forms of IT in underpinning the development process is often under-valued."

The role of advanced telecommunication systems and other forms of IT in underpinning the development process is often under-valued. Clearly, they will not solve problems of disease, contaminated water supplies, inadequate sanitation or overcrowded transportation systems on their own, but experience shows that they can help hard-pressed urban authorities identify and tackle priority problems. In Niger, for example, the Ministry of Public Works and Town Planning has successfully used satellite imagery to plan and manage highway maintenance.[83]

The role of IT in informing the management of the *global commons,* including the atmosphere, oceans and seas, has already been discussed in Chapter III. Clearly, data from sensing instruments and data processors will continue to assist our understanding of, and response to, problems caused by the El Niño phenomenon, global carbon dioxide build-up, acid rain, or variations in the stratospheric ozone layer.

The commercial value of Third World applications of IT in the management of the global commons remain somewhat small, however. Also, while the newly industrializing countries may be better able to afford the relevant technologies, the poorer nations often have the most urgent need. As a result, many programs will need to be initiated, coordinated and funded by international aid and development agencies. Nonetheless, it is increasingly apparent that new market opportunities will be created by the drive for sustainable development. To get a better idea of how the more forward-thinking IT companies view this market potential, Chapter V focuses on the world's largest computer company, IBM.

V.

IBM: A CASE STUDY

Despite its recent problems, IBM still could be the world's largest company by the year 2000. "I don't care who you are in this industry," said IBM chairman John Akers in 1986. "No-one can compete with the IBM company." IBM's revenues scarcely rose in 1986, the company's worst showing in 40 years, but they still amounted to $50 billion. Profits were down, for the second year in a row, but IBM still earned nearly $5 billion.[84] According to some sources, IBM's profits still represent about 70 percent of the entire information industry's profits.[85]

"Despite its recent problems, IBM still could be the world's largest company by the year 2000."

Certainly no other computer company can match the sheer scale of IBM's investment program: in 1985, for example, IBM spent $4.7 billion on research, development and engineering, with a further $3 billion going on automating its factories. As *Business Week* noted, IBM invested $1 billion more that year than its closest competitor, Digital Equipment (DEC), reported in total turnover.[86]

Arguably the most sucessful company in the world, IBM has been the subject of long-running controversy. Contrasting views of the reasons for the company's success have been outlined in two recent books: David Mercer's *IBM: How the World's Most Successful Corporation is Managed*[87] and Richard Thomas DeLamarter's *Big Blue: IBM's Use and Abuse of Power.*[88] Mercer, a long-time IBM executive, stresses the importance of a management style that supports individualism and innovation, demands the highest ethical standards, and tolerates continual change. DeLamarter, on the other hand, concludes that IBM has succeeded because it has always been a ruthless monopolist.

If you subscribe to DeLamarter's view, IBM's growing interest in expanding into telecommunications, banking, and robotics is alarming. If, on the other hand, you are looking for a single corporate "lens" through which to focus on the computer industry's potential contribution to sustainable development, IBM is the obvious choice. First, and most obviously, it is the world's largest computer company, with a research and development budget to match. Second, it has Scientific Centers (see page 64) located in a number of countries around the world which, among other things, are developing hardware and software with relevant applications in mind. Third, in addition to its traditional mainframe, minicomputer, personal computer, and software markets, the company has been expanding into telecommunications, satellites, robotics, electronic information services, and artificial intelligence work. And, fourth, IBM also happens to be one of the few computer companies with any real sense of what the term "sustainable development" actually means.[89]

In Britain, for example, IBM UK has helped fund the Centre for Economic and Environmental Development (CEED), set up to implement the *Conservation and Development Programme for the UK*[90]—prepared in response to the *World Conservation Strategy.* The company's chief executive, Tony Cleaver, is on CEED's board and the computer company has incorporated sustainable development into its corporate and social responsibility programs.

A review of the company's "sustainability" carried out for CEED by Brian Johnson in 1985 concluded that IBM had made enormous progress, but was equivocal on whether the sort of compound annual growth IBM is aiming for squares with sustainable development.[91] The company's U.S. revenues grew at an average annual rate of 14.8 percent between 1975 and 1985. Adjusted for inflation, this was real growth of 8.2 percent, compared with 2.9 percent for U.S. GNP. This represented a growth rate nearly three times as fast as the economy as a whole, and between 1982 and 1984 IBM's U.S. revenues grew, in real terms, seven times as fast as the economy.

To reassure those who felt such growth rates were environmentally unsustainable, IBM developed its own internal environmental management systems. Indeed, the company has formally recognized the importance of environmental protection for many years. When it published its corporate environmental protection policy in 1973, it noted that "IBM is not in a business which creates severe pollution problems," but nevertheless committed itself to: (1) meet or exceed all applicable government regulations, in any location; (2) establish stringent standards of its own where such regulations do not exist; (3) attempt to use nonpolluting technologies and to minimize energy consumption in the design of products and processes; (4) minimize dependence on end-of-the-pipe waste treatment by developing techniques to recover and reuse air, water and materials; and (5) to "assist government and other industries in developing solutions to environmental problems when appropriate opportunities present themselves and IBM's experience and knowledge may be helpful."[92]

Sustainable development is clearly one of those areas, and IBM has considerable analytical skills, technology, and financial clout to offer. The company's growing interest in sustainable development may help bring into focus some of the fairly diverse environmental strands in its research and development programs over the years, but it is still too early to say what the overall impact of this new orientation might be. Meanwhile, too, some investors have questioned whether IBM has been focusing enough on its own commercial sustainability. The computer market slumped in 1985, first in the United States, and then in Europe. While demand for minicomputers and microcomputers continued to grow reasonably rapidly, the demand for mainframes has plummeted. As the company's ex-treasurer, Jon W. Rotenstreich, recalled, "It was like falling off a damn cliff."

Apart from the general slow-down in demand for computers, other factors were at work. For one thing, customers now wanted to network their computers, and IBM was operating at least nine major computer "architectures," many of which were incompatible. Second, IBM's share of the microcomputer market had fallen from a peak of around 70 percent when the IBM PC was first launched to around 27 percent, and seemed destined to fall further: hundreds of smaller companies were offering similar systems at half the price, and IBM was behind the field in introducing new generations of microcomputers based on Intel's powerful 386 chip. Third, IBM recognized that it must pay more attention to what the customer actually wants—witness the new consensus that the company must "focus on the customer's problem, get inside the customer's head"[93] and do more to boost innovation. Finally, the company's diversification attempts have not always met with success. Its first major telecommunications venture, a U.S. business communications network called Satellite Business Systems (SBS), began operations in 1981 but failed to make money. IBM sold out in 1986 to MCI. (Inmet, a joint venture with Merrill-Lynch in the electronic financial information market, was also scrapped.)

IBM recognizes that if it wants to meet its ambitious targets for growth, it is going to have to push into new and unfamiliar markets—where the risks are high and its huge size may not be an asset.[94] Managing economic development in such a way that environmental resources are conserved and renewed is a highly information-intensive business, although not yet highly developed in commercial terms. So what efforts has IBM been making to "get inside the customer's head" in the sustainable development field?

The answer is that, given the field's early state of development, IBM has made a significant effort. As far back as the early 1970s, for example, the company was involved in urban pollution and air quality modeling projects focusing not only on U.S. cities like St Louis and New York, but also on Third World cities like Mexico City.[95] Its computer modeling work helped to cut the time needed for many of the calculations needed in air quality modeling by several orders of magnitude.

While IBM's Scientific Centers in the U.S. pulled out of pollution modeling in the 1970s, IBM Italy is still carrying out a fair amount of modeling work focusing on the air quality impact of a power station and on the management of groundwater resources in the Friuli-Venezia Giulia region. The groundwater work is using the IBM Hacienda (Highly Advanced Color Image Enhancement and Display Architecture) system to display the characteristics of the aquifers under study.

The role of the Scientific Centers, explained Dr. Horace Flatt of the Palo Alto Scientific Center, is to "look at future customer requirements, with the aim of developing the appropriate equipment. We represent a long-term IBM investment in broadening computer markets. We look at the state of the technology and ask: what do we need to do to get from here to there?"

In the environmental field, he noted, "faster circuits and bigger memories are driving applications forward, but many environmental monitoring applications need extraordinarily fast data input rates from the sensors." Where IBM has tackled environmental problems, whether in air quality modeling or remote sensing, it has selected them as challenging technical problems whose solution could help spur the development of hardware and software, rather than as a gesture towards social responsibility.

Remote sensing, said Dr. Flatt, "is an attractive application from a computer manufacturer's point of view. It uses an awful lot of computer time to process the data, and it has some very interesting display requirements, which tend to force up the cost of using it." The problem, he cautioned, is that "it costs a great deal of money, and most of the identifiable users don't have a lot of money."

Although IBM has made some important contributions in remote sensing, however, it has been switching effort into other applications of its image processing techniques. "I think that in some ways our decision to leave remote sensing was because we felt that the field had been oversold," said Antonio Santisteban, director of scientific programs at IBM's Madrid Scientific Center. "Remote sensing, especially in the developing countries, will have to wait for the price of digital processing to drop."

But the price of the necessary computers is falling all the time. Dr. Flatt noted that the cost of a high resolution image processing system, connected with a larger host system would be around $150,000, but this price-tag is likely to fall to $50-$75,000 over the next few years.

The Palo Alto Center, for example, has developed an experimental image processing system based on the IBM PC, providing interactive multispectral image processing and image analysis capability for small images. Because the data processing load is directly proportional to the *square* of the image's dimensions, IBM scientists Joseph Myers and Ralph Bernstein point out, "it is not surprising that the technology has been restricted in the past to only those with access to large computers or special-purpose image processing systems."[96] Their results show that, while mainframes and minicomputers will still be needed for the more complicated number-crunching jobs, microcomputers could soon handle many important tasks which were previously outside their range. *A Guide to Software for Developing Countries*[97] has been compiled by IBM, together with a report, *IBM in Africa,*[98] outlining some of the development-related applications of the company's technology. The Software Guide catalogues programs under four broad headings: agriculture, economic and social resources, physical infrastructure, and administration, describing programs such as the FAO's Farm Analysis Package, UNCTAD's Debt Monitoring System, and USAID's Demographic Projection Model for Microcomputers.

Ultimately, indeed, IBM's specific work on environmental projects may have less of an impact on

39

"Ultimately, IBM's specific work on environmental projects may have less of an impact on the prospect for sustainable development than its mainstream role in reinforcing the worldwide trend towards cheaper computing power."

the prospect for sustainable development than its mainstream role in reinforcing the worldwide trend towards cheaper computing power. IBM's chief scientist, Ralph E. Gomery, predicted in 1986 that the future for the computer industry is likely to be pretty much like the past, with a continuing 20 percent year-on-year decline in the cost of computing. Cheaper computers will unlock new environmental applications of IT, some of which are touched on in Chapters III and IV.

It is worth noting, however, that while IBM PCs may be used at the Madrid Scientific Center to model oil pollution in the North Sea, for example, in Venezuela they also are being used to identify mineral deposits in the Orinoco region, including forested areas. Computers provide tools which can help exploit environmental resources or protect them, with the exact balance of advantage depending on the political context within which they are used.

VI.

HARNESSING THE POTENTIAL

Information technology, it has been said, is a "heartland" technology, like steam, electricity, or the internal combustion engine. It has the potential to radically change the world we live in: the way we learn, the way we work, the way we communicate, even the way we think.

> *"Typically, when policy-makers think of the 'sustainability' of the U.S. IT industry, they think in terms of foreign competition; trade barriers and impending trade deficits; the impact of computerization on the structure of the economy or on employment; the strategic implications of computer sales to Third World countries—or the potential for the accidental or intentional invasion of privacy."*

Typically, when policy-makers think of the "sustainability" of the U.S. IT industry, they think in terms of foreign competition; trade barriers and impending trade deficits; the impact of computerization on the structure of the economy or on employment; the strategic implications of computer sales to Third World countries—or the potential for the accidental or intentional invasion of privacy. As first chips and then computers have become internationally traded commodities and the pressure on the U.S. chip and computer industries has intensified, the agendas of such key government agencies as the Office of Technology Assessment and such industry associations as AFIPS (see page 54) have very much focused on these "mainstream" issues. But there is clearly a willingness to address new issues that might significantly affect the industry's future prospects.

As far as the Third World is concerned, it is clear that IT *will* have a profound impact. What is not yet clear is where and when the impact will be felt, and in what way. Some analysts believe that these emerging technologies afford the developing countries an opportunity to "leapfrog" the many steps taken by industrialized nations in building their IT industries. Others suggest that the labor-saving aspects of IT will enable manufacturing industries to move back from the Third World into the industrialized countries, a reversal of the de-industrialization trend of recent decades.[99] Overall, the conclusion must be that the effects of these new technologies will be complex, both direct and indirect, and spread unevenly among different countries, regions, and industrial sectors.

Current trends in IT include an accelerating pace of development, enhanced connectivity both within and between systems, and a growing convergence between different types of information collection, communication, processing, storage, and presentation technologies. While hardware developments are critically important, however, the contribution of software should not be overlooked. Jim Manzi, Chief Executive Officer of Lotus Development Corporation, has noted that computer users all-too-frequently suffer from "technology indigestion."[100] These problems are particularly acute for Third World users.

The job of software producers, according to Manzi, is "to help the customer through this technological mess." The company's best-selling Lotus 1-2-3 spreadsheet has been used, for example, to study the effects of acid precipitation. And Lotus has also developed Measure, a software package that automates data entry into Lotus 1-2-3 from remote sensors. At least one journal, *Environmental Software*, is now reporting on some of the emerging environmental applications.

Whether or not it currently recognizes the fact, the emerging IT business has a major stake in the sustainable development of the world's natural resources. Adverse environmental trends which may undermine the fortunes of customers, whether they be corporations or countries, ought to cause concern. Very few companies have even begun to consider the commercial potential of sustainable development, but IBM's interest may spur other organizations in the same general direction.

"Whether or not it currently recognizes the fact, the emerging IT business has a major stake in the sustainable development of the world's natural resources."

Unlike the biotechnology industry (the subject of the first report in this series), the IT industry has not—apart from isolated pollution controversies—been a target of sustained environmental lobbying.[101] There may be some disadvantage in this, in that the IT industry has not been forced to address the environmental agenda in the same way as the biotechnology industry has. But it is clear that some

of those working in this industry, which shows every sign of becoming the world's largest industrial sector by the early years of the 21st century, are aware of their stake in the global environment.

"We recognize that the technology of our industry, information technology, cannot replace the natural ecosystems that are so badly threatened," explained Tony Cleaver, Chief Executive of IBM UK, "but we do believe that technology, properly used, can be a key ingredient in managing the resources of this planet more sensibly and *that*, at the end of the day, is what sustainable development is all about."

"The market mismatch between Third World needs and the predominantly First World locus of the IT industries means that most existing programs are merely scratching the surface of the problems and the potential of IT in sustainable development."

There is a growing need to build bridges between the new industries whose activities are sketched in this brief report and those who promote sustainable development, both in the industrialized regions and in the Third World. Various initiatives are already under way, as previous Chapters have indicated, but the market mismatch between Third World needs and the predominantly First World locus of the IT industries means that most existing programs are merely scratching the surface of the problems and, it should be noted, the potential of IT in sustainable development.

The pace of technical developments in remote sensing, telecommunications and information technologies will continue to accelerate. So it is increasingly important, particularly given the levels of investment likely to be involved, that industrial and environmental interests exchange views early on in the development process, before new—and perhaps unsustainable—patterns of development are effectively set in concrete or silicon.

Unfortunately, environmentally aware senior industrialists like Tony Cleaver remain very much the exception rather than the rule in the IT industry. The question, then, is whether such companies as IBM, Apple, Cray, or DEC are likely to be natural allies in the drive for more sustainable forms of development? The evidence suggests a somewhat cloudy forecast. Indeed, Apple's recent history provides a telling example of the way in which business pressures and priorities can constrain the more expansive goals of even fairly wealthy corporations (see page 55). And an even more striking example is provided by Control Data Corporation (CDC), profiled on page 59.

For many years, CDC's goal was to apply its technological, financial, and human resources "to address major unmet needs of society." This strategy took the company into areas related to sustainable development, from energy conservation and the development of renewable energy sources (e.g. CDC's PLATO/BLAST energy audit program and TECH-NOTECH database), through rural development programs designed to sidestep the water shortages, soil erosion, and other problems associated with large-scale farms, to urban revitalization projects. The idea, CDC stressed, was not philanthropy but the opening up of new business opportunities.[102]

Working with the Denver Research Institute in the early 1980s, CDC also built up the DEVELOP database and information service, aimed at users in developing agricultural and industrial economies. The idea was to provide up-to-date information, for example, on alternative energy systems, small-farm agriculture, small-scale enterprise, and appropriate technologies.

Ultimately, however, the sheer diversity of projects being pursued helped pull CDC into the red. The company's new president, Robert M. Price, was forced to scale down, write off, or sell many of the pioneering social projects that CDC had helped build up. DEVELOP was one of the casualties. "We don't have any visions," Price explained later. "We have a damn job to do, and that's to make the company profitable." The shift in strategy was summed up by one employee in the following words: "We used to be a culture-driven business. Now we're a business-driven culture."[103]

CDC still aims to apply its technologies where they can *profitably* and *productively* address unmet needs, however. "Some, but certainly not all, unmet needs represent major opportunities for (CDC's) technology and projects," as Price put it. "As with all our businesses, we discriminate as to time-frame, the level of investment required, and the degree of potential among those opportunities. Most important, we operate business ventures, not 'social programs.' "

Would appropriate Third World applications meet these criteria? That depends on which markets a company was trying to break into. CDC, in fact, has been helping India build up its own computer industry with a novel eight-year agreement covering the transfer of the entire manufacturing process for its medium-powered Cyber 180 computers to the Electronics Corporation of India. The resulting machines are expected to find applications in engineering and scientific analysis, banking, transport, education, and research.[104] As a result, if India decided to apply such systems to environmental management problems, it could do so that much more easily.

But, while the sustainable development programs are likely to be fairly information-intensive, with a need to integrate growing volumes of data on population trends, economic activity, and the health of natural systems, it does not yet represent a major market opportunity of the type that CDC would find irresistible. It is worth remembering, however, that it may offer an entry point into tomorrow's growth markets in the newly industrializing countries.

Many Third World countries feel that IT may offer them their only chance of closing the gap between their own economies and those of the industrialized nations. But the fact remains that many analysts are convinced that the gap between North and South will continue to widen. "The Third World is likely to remain at the economic periphery for a long time to come, both from the perspective of the entire electronics market, and in terms of specialized telecommunications systems," concluded *South* magazine in a recent special report on IT.[105] Nonetheless, 10-20 percent annual growth rates are expected in communication and computer markets in such countries as India, Brazil and Singapore. The Third World share of world IT markets is forecast to grow from 9 percent to 14 percent by 1990, even though the absolute gap between North and South, measured in billions of dollars, will be greater than ever before.

Further, a number of diverse problems have emerged as some developing countries have attempted to develop their own IT industries. India and Brazil, for example, introduced new laws to promote their indigenous computer industries. However, in India, IBM pulled out when the government asked for 51 percent of IBM India,[106] while in Brazil, Microsoft, unable to sell its MS/DOS software because of sanctions blocking U.S. computer sales there, saw the business go to a Brazilian company which admitted copying Microsoft's internal coding.

There are enormous differences, too, in the capacity of different Third World countries to absorb the new technologies. Some are placing growing stress on computer literacy: Singapore, for example, set up its National Computer Board in 1981 to establish the country as a computer software and services center for the 1990s. Sri Lanka, reports Jack Fritz of the National Academy of Sciences (NAS), "only had a few computers at the end of the 1970s and no formal computer training programs. Now there are an estimated 2,000 to 3,000 small microcomputers, 500 to 600 large microcomputers, three university computer science departments, microcomputer laboratories in eight of the nine universities and in more than 100 secondary schools."[107]

The U.S. Agency for International Development (AID) has been collaborating with the NAS on a series of international conferences to explore how computers can assist development. The first took place in Colombo, Sri Lanka's capital, in 1984. It focused on such applications as farm management, population planning, and the design of water and sanitation systems. The Mexico symposium, held in 1985, looked at the use of microcomputers in education and training, and concluded that a dearth of suitable software is limiting the spread of computers. The third conference, held in Portugal late in 1986, covered the role of artificial intelligence, mass storage and telecommunications in development.

Overall, the experience of the industrialized countries suggests that IT has enormous potential to help developing countries address some of their most urgent problems, whether they be in crop production, irrigation scheduling, livestock management, rural health care or environmental protection. But there are enormous barriers to the spread of computers in the Third World, let alone computers dedicated

to environmental and resource management applications.[108] In the telecommunication field, for example, many manufacturers have been withdrawing from the under-developed and under-funded markets of the Third World. Best placed to penetrate cash-poor countries, according to Lars Ramqvist of Sweden's Ericcson, are the Japanese. Supplier credit and counter-purchase deals are widely used by Japanese companies, indeed one trading company recently paid for a shipment of Venezuelan iron ore with telephones.

"Supplier credit and counter-purchase deals are widely used by Japanese companies, indeed one trading company recently paid for a shipment of Venezuelan iron ore with telephones."

Wherever the technology comes from, our planet is shrinking. Growing numbers of people, using ever-larger volumes of raw materials and energy to support ever-more complex lifestyles, are putting natural systems under ever-greater pressure. IT offers a number of solution-oriented technologies which can help to identify, monitor and manage emerging environmental trends. If the resulting information is to be of value in the Third World context, however, the emphasis must be on accessibility and cost-effectiveness. The trend toward privatization and reliance on market mechanisms in the remote sensing business will place a brake on some desirable applications of IT.

"The non-commercial or 'soft values' (that is, those related to improving the quality of life and the environment) tend to be the first ones compromised when remotely-sensed Earth resources data costs are increased," noted Stan Aronoff of Canada's Dipix Systems.[109] The effect of rising prices, he predicted, would be to shift the distribution of applications "away from resource conservation and environmental monitoring in favor of resource development." Indeed, as another commentator put it, "commercialization of data collection and dissemination in advance of developing a market demand for those data is the same as charging $1.30 per gallon for gasoline before there were automobiles to use it in. At those prices who would have learned to drive?"[110]

Ultimately, the policy context in which such data are analyzed is likely to be the critical factor determining the extent to which all this activity furthers environmental sustainability. With the U.S. space industry at a major crossroads, and the semiconductor and computer industries in a protectionist mood, the political climate is perhaps not particularly favorable for new international initiatives, such as the "ENVIROSAT" concept discussed in Chapter III.

"Whatever happens to ENVIRO-SAT, it is extremely unlikely that the development and deployment of appropriate informations technology will keep pace with Third World needs if left to market pressures alone."

Whatever happens to ENVIROSAT, it is extremely unlikely that the development and deployment of appropriate informations technology will keep pace with Third World needs if left to market pressures alone. It hardly needs stressing that many of the most pressing Third World needs simply do not register in the international market-place. As a result, commentators have asked whether the future could see a new form of colonialism, based not on the control of international sea routes but on the possession of bigger and better information processing and telecommunication systems? Certainly, the pace at which IT has been developed and adopted has been far from uniform around the globe. It has been said that 90 percent of the activity has happened in 10 percent of the countries.

A number of international agencies have initiated a number of projects designed to promote the diffusion of IT in developing countries. UNICEF, for example, launched the pilot phase of its UNET network. The initial objective was to ease communication between UNICEF's own offices around the world and fund-raising committees in North America and Europe, but the potential for using the network for worldwide communication soon became obvious.[111] Once the system is in place, anyone with access to a computer, modem and telephone line can plug into the network. Although there have been technical gremlins, many of them caused by the need to integrate new technologies with older telephone switching systems, the overall assessment is that, in the short term, UNET is helping improve the links between UNICEF and its field staffs. In the longer term, the potential for expanding the net is enormous.

Some innovative IT programs are explicitly linked to trade. When the Department of State supported a "distance learning" joint venture between Tele-Teaching International, Inc., and China's State Education Commission, for example, the hope was that exposing the Chinese to American IT systems would generate over $250 million in business contracts involving the sale of hardware, transmission charges, provision of instructional services and consulting fees.

Third World IT markets may well take time to develop, however. The experience of countries like Haiti, where the Service Nationale des Endemies Majeures (SNEM) has been trying to automate the epidemiological and administrative elements of its malaria control programs, suggests caution. The Data General DG-10 which formed the heart of the chosen system was delivered without software—and when the problem was eventually sorted out, the software which had been ordered was not compatible with the computer's operating system.[112] "Training, the essential ingredient in computer applications," noted Dr Vishnu-Priya Sneller of USAID, "had been overlooked." There is a real need for computer training programs designed to outlast the departure of foreign trainers. The work carried out in Sri Lanka (see Chapter IV) by Mohan Munasinghe, energy advisor to the World Bank and President of the Sri Lanka Energy Managers Association, provides a useful model of what can be achieved.

"Computer training programs must be designed to outlast the departure of foreign trainers."

One U.S. firm that specializes in advising on suitable computer systems for Third World operations is Thunder & Associates, based in Alexandria, Virginia. "Like any successful technology, the microcomputer is 'appropriate' if it is used in the right setting, for the right purposes, and with the right infrastructure to support it," stresses Noel Berge, the firm's president and author of *Microcomputers in*

Development: A Manager's Guide.[113] The problems facing developing countries in taking computer technology on board are the same as those originally faced by the developed countries: a lack of trained personnel, non-standardized hardware and software components, unstable power supplies, and all the other problems with working on the technological frontier.

Although the Apple Macintosh is a rugged, dependable computer, with reasonable support facilities around the world, says Lizbeth Shay, also of Thunder & Associates, its reliance on a proprietary operating system has limited its compatibility with other systems.[114] The most widely used microcomputer in development-related applications is the IBM PC (XT and AT), followed by IBM compatibles made by such U.S. firms as Compaq, Radio Shack, or Kaypro, or by foreign (particularly Asian) firms. Apple, however, is now moving towards IBM compatibility.

To help Third World PC users identify suitable sources of software, IBM produced *The Guide to Software for Developing Countries* (see Chapter V). The software, from sources such as the U.N. system, the World Bank and the International Rice Research Institute, covers such wide-ranging applications as the management of catfish and timber production, demographic modelling, water supply, and urban planning. It also identified many relevant catalogs, reports, and newsletters.[115]

A similar guide to the environmental and resource management software now available for Third World users would be a valuable tool. A useful step in that direction was *Managing a Nation: The Software Source Book* published by the Global Studies Center,[116] although this reference work largely addresses U.S. needs and applications. Among the software reviewed were programs or models focusing on urban and rural development, energy and water supply, agriculture, forests, population, environmental management (all of which, perhaps coincidentally, had been developed for Apple computers), and transportation.

The original U.S. lead in software development has been eroded by international competition, particularly from Europe and, increasingly, Japan. But, although it is certainly needed, software adapted to Third World needs is unlikely to be a top priority in any of these competing regions. So will the developing countries now be able to get in on the act? Robert Schware, an international affairs fellow of the Council on Foreign Relations considered this question and concluded that there are a number of significant barriers to their participation in this new industry.[117] Their low wage scales and large labor pools, which would seem to be an advantage, will not help as software development becomes less labor-intensive—because of the increasing automation of software development.

According to Schware, those countries (for example, Brazil, India, Sri Lanka, and Malaysia) that have already built—or are in a position to build—the technological infrastructure necessary for software development will be better placed to attract foreign investment. Companies like Texas Instruments (TI) and Citibank have set up their first overseas software design centers in India, because of the country's cheap, talented labor. TI's $5 million installation in Bangalore is linked by satellite to the company's Dallas headquarters, so that TI's Indian engineers can export their software literally at the touch of a button. This building strength in software helps Indian hardware producers, too. Tata Unisys, formerly Tata Burroughs, began building computers in India in the late 1970s and added microcomputers to its range in 1987.

Other less developed countries, such as Syria, Tanzania, or Zambia, are unlikely to build a major software export capability, but might develop appropriate software for local or regional markets if appropriate government policies are adopted. Overall, however, the pace of software development is so rapid that the potential returns from investment by Third World nations would need to be very carefully evaluated.

For Third World IT markets to take off, there will need to be strong domestic interest in computers and computing. The role of small companies, such as PCC (see page 69), in increasing overall computer literacy should not be under-estimated. PCC's activities have been mainly focused in the United States to date, but the enthusiasm and ability of such organizations to focus on local needs, harnessing local resources, can make them particularly effective agents of change. They hold the potential for a genuine "bottom-up" approach to development, rather than the "top-down" approaches often favored by government agencies and large companies.

The adverse impact of rising prices for remote sensed data will depend on a number of factors. But one is certainly cost: the magnetic tape version of a thematic mapper Landsat image representing around 185 km by 170 km of the Earth's surface, was priced at $3,300 in 1986, more than the cost of many microcomputers. How fast the cost of equipment, microcomputers included, falls is critical. So is the extent to which the U.S. "value-added" industry manages to enhance the value and "user-friendliness" of remote sensed data. Given the important role that value-added companies are playing, organizations involved in sustainable development should keep a close eye on their activities.

Meanwhile, the high cost of satellite data makes the establishment of developing country value-added industries difficult, though innovative approaches are being developed. In Kenya, for example, small organizations—such as Ecosystems Ltd., based in Nairobi—have built up impressive renewable resource GIS systems. Using microcomputer equipment costing around $6,000, and standard Dbase III software, the company has used remotely sensed data from light aircraft to beat the high cost of satellite data and meet the demand for a higher resolution than is afforded by Landsat images (30 x 30 meters). While satellite data may be extremely useful for First World agricultural applications, the difference between starvation and plenty in many Third World countries is often a question of the productivity of small fields and plots, many of which are currently invisible to satellite sensors.

While satellite systems have a unique ability to provide a synoptic view of the globe, developing countries will require data appropriate to their particular needs. As Ken Craib of Resource Development Associates International (RDA: see page 69) notes of their developing country clients—"they don't want space data, they want data. If they can get the information they need by riding into an area on a mule, then they don't need remote sensing data. One early firm, ERA, collapsed because they failed to realize this."

Pitfalls notwithstanding, enthusiasm for satellites in such countries is considerable. Professor Odingo of the University of Nairobi, who has been lobbying for the establishment of an African Remote Sensing Society, is convinced that satellite technology will enable developing countries to leapfrog the institutional and funding hurdles slowing the deployment of modern information and communication technologies.[118]

Various initiatives are afoot to pull down the cost of satellite communication and data. One is the low-cost satellite communications system that the Volunteers in Technical Assistance (VITA), based in Arlington, Virginia, have been developing. Dennis Meadows, co-author of *The Limits to Growth* study, has also been using cheap ($200-$400) lap-top computers in his Third World training programs.

But rising prices are inevitable in some areas—and will discourage the use of remote sensing data for some environmental applications. The basic problem is that market mechanisms are unlikely to meet the needs of the poorer countries in a reasonable time-scale. Even in the developed countries, as the Industry Advisory Group of NOAA put it in 1982, "the market for remote sensing data has not developed to a point where commercial viability can be demonstrated within the next 10 years." The picture has not changed significantly during the intervening years.

Remote sensing, commented Karl-Heinz Szekielda of the United Nations Department of Technical Co-operation, "is an additional tool, with a relatively marginal impact. There are no big market applications. If anything, the technology is going down, not up. The user community is shrinking, because prices have gone up by several orders of magnitude."

But satellite remote sensing data are unique. As Chapter III explains, they offer a synoptic perspective of our planet, which makes them a driving force behind some of the major emerging global research programs. Given the proportion of such data which are generated for military use, and consequently subject to tight security requirements, there is an urgent need to work out ways of providing access to relevant parts of the resulting data sets for *bona fide* environmental researchers.

As large-scale research programs using such data grow in importance, so the volume of data that will need to be handled will grow almost exponentially. The data handling and storage problems facing organizations like NOAA, which collect worldwide environmental data, are considerable, although advanced computing technologies, including super-computing, geographical information systems, and expert systems, will increasingly provide the necessary tools for managing such data flows. NOAA makes such data available through service centers operated by NESDIS. But it is still far from clear how such technologies and data can be made both accessible to—and user-friendly for—potential Third World users working in environmental management and sustainable development.

> *"It is still unclear how advanced computing technologies and data can be made both accessible to—and user-friendly for—potential Third World users working in environmental management and sustainable development."*

Part of the problem is that many international organizations are themselves fragmented, both administratively and geographically. Networks like UNICEF's UNET provide the means for ensuring a more coordinated response to complex development problems, if there is a will to use them. They also provide a feedback mechanism, so the progress of development projects can be monitored long after the consultants or extension workers have moved on to other tasks. The importance of system compatibility cannot be overstressed, however. Even IBM, whose system languages have become global standards, has produced computer ranges which are mutually incompatible. The potential for creating high-tech Towers of Babel is only too clear.

The leading aid agencies need no convincing that appropriate technology transfers should be promoted, however. "Problems like tropical deforestation are often caused by millions of small farmers mismanaging their environment," notes Dr Charles Paul of USAID. "So it's important to ask what sort of difference the right sort of information might make at that level. The Peace Corps have shown that it really can make a difference. In remote villages, microcomputers have an important role to play as an integrated planning tool. Whether you are talking about how long it takes the women to fetch water or the rate of soil erosion, the potential of a 'People's GIS' could be considerable."[119]

While good work can be done without IT, those who make or shape major national or international policy decisions need to know that these emerging technologies can help us move toward much better informed styles of environmental management, be they local, regional, national or global in scope. "Our objective," as World Bank President Barber Conable put it in an address to WRI in 1987, "is a sort of natural resources balance sheet, a coherent planning instrument for better management." Inevitably, IT would play a critical role in compiling and updating this balance sheet. "What I seek from data—much of which is already at hand," Mr Conable continued, "is a composite inventory of environmental assets and liabilities. With such a planning instrument, we could move toward establishing the value of those priceless resources—topsoil and grass cover, water and drainage, human skills and traditional lifestyles—we too often treat as worthless."[120]

> *"In light of growing international competition it is not clear that U.S. companies are doing enough to keep abreast of market trends in the less developed countries of the world. Companies that focus exclusively on the bottom line, quarter by quarter, may well miss significant long-term opportunities."*

The development aid agencies, industry associations, manufacturing and service companies, and non-governmental organizations all have important roles in ensuring that their relevant constituencies understand the role that IT could potentially play in supporting sustainable development. The experience of Control Data suggests that a strategic perspective—coupled with careful attention to the bottom line—is as essential in this area of business as in any other. But in the light of growing international competition it is not clear that U.S. companies are doing enough to keep abreast of market trends in the less developed countries of the world. Companies that focus exclusively on the bottom line, quarter by quarter, may well miss significant long-term opportunities.

As IBM has put it, the IT tools needed for sustainable development are increasingly available. The key, as AT&T Executive Vice President John S. Mayo explained the concept of "information productivity," will be "getting the right information to the right place at the right time." The next step must be to form practical partnerships among government agencies, industry associations, interested hardware and software suppliers, and organizations working toward sustainable development. The goal: to refine existing products for Third World use, and to identify the emerging market opportunities for new generations of products and services tailor-made for sustainable development.

John Elkington is a Director of SustainAbility, Ltd., and of John Elkington Associates, both based in London, England. Previously, he was Managing Director of Environmental Data Services, Ltd. The main focus of his work is on the business side of sustainable development. He wrote an earlier report in this series, *Double Dividends? U.S. Biotechnology and Third World Development* (1986). **Jonathan Shopley** is an Associate at John Elkington Associates.

BUILDING BRIDGES

It would be easy to oversell the near-term potential of information technology in general—and of satellite remote sensing in particular—for promoting sustainable development. Although the IT business is clearly destined to become one of the pivotal sectors of the global economy, a very small proportion of IT applications to date have had anything to do with environmental protection or the sustainable development of natural resources.

- There is a clear challenge for environmental and sustainable development organizations to ensure that the rapid growth of the IT sector is matched by a considerable expansion in the number of applications designed to support sustainable development.

From the point of view of the IT industry, most existing—and many potential—applications in this field represent "niche," or small-scale, markets. At the same time, however, environmental protection and sustainable development are extraordinarily "information-hungry."

- Sustainable development programs' need for a great deal of computing power represents a useful building-block in any attempt to create bridges between the industry and organizations active in this field.

Despite the rapid fall in hardware costs, even relatively modest IT systems are still outside the means of many potential Third World users. At the same time, the cost of data, particularly remote sensing data, has been rising, largely driven by the commercialization of activities that were previously heavily government-sponsored. Meanwhile, international competition in the IT business is so intense that most hardware and software companies are unlikely to invest in apparently remote and long-term areas such as sustainable development.

- Given these constraints, it is even more important to ensure that IT industry initiatives in support of sustainable development in Third World countries are not unnecessarily compromised by trade regulations or by the commercialization of sources of technology and data previously wholly in the public sector.

The bulk of current IT expenditure in the United States is invested in defense applications. In this context, it is worth noting that a key objective of sustainable development is to secure the future of environmental resources essential to national and international security.

- While critics of defense spending might advocate the beating of IT "swords" into "ploughshares," a number of immediate practical steps would ensure that some of this spending is diverted to sustainable development programs. For example, can the enormous databanks built up by—and on behalf of—the U.S. Department of Defense be selectively opened up to organizations interested in this broader concept of security?

It would be ill-advised to urge the transfer of U.S. information technologies to Third World users in the hope that the result would be sustainable development. To ensure that these powerful new tools are used to support the sustainable management of environmental resources, both governmental and non-governmental organizations have a key role targetting and professionally managing IT technology transfers and investment. If poorly managed, technology transfer may actually increase Third World dependency on the industrialized nations. Without sufficient training, and either an indigenous manufacturing base or the means with which to buy further equipment and services, such initiatives will achieve little.

- Clearly, though the IT industry generally supplies the technology, national and international government agencies must help build up the indigenous capabilities needed in Third World countries to sustain the use of such technology.

Part of a WRI series on the implications and applications of emerging technologies, *The Shrinking Planet* has sketched out key trends in the IT industry. It has also focused on a number of applications that could promote sustainable development. The object has been to broaden understanding of its potential contribution.

- To explore and develop this potential, a consortium of interests needs to be brought together, including industry representatives, national and international development agencies, and non-governmental organizations with an interest in sustainable development. If First World enthusiasm and capabilities are to engage with Third World realities and needs—as they must, the early involvement of potential Third World partners is crucial.

GLOSSARY OF ACRONYMS

ABLE   NASA's Amazon Boundary Layer Experiment

AFIPS   American Federation of Information Processing Societies

AI   Artificial Intelligence

AID   United States Agency for International Development—also US AID

ARCO   Atlantic Richfield Company (US)

ARTEMIS   System Definition of Africa Real-Time Environmental Monitoring Using Imaging Satellites (FAO project)

AT&T   American Telephone and Telegraph Company

AVHRR   Advanced Very High Resolution (remote sensing) Radiometer

CAD   Computer Aided Design

CAE   Computer Aided Engineering

CAM   Computer Aided Manufacturing

CBEMA   Computer & Business Manufacturers' Association (US)

CD   Compact (laser optics) Disk

CDC   Control Data Corporation (US)

CD-ROM   Compact Disk Read Only (computer) Memory

CEED   Centre for Economic and Environmental Development (UK)

CITES   IUCN's Convention on Trade in Endangered Species of wild fauna and flora

CLEARS   Cornell Laboratory for Environmental Applications of Remote Sensing

CMC   Conservation Monitoring Centre of the IUCN (UK)

CNES   French National Space Center

COBOL   Common Business Oriented (computer) Language

CZCS   Coastal Zone Color (remote sensing) Scanner

DEC   Digital Equipment Corporation (US)

EC   European Community

EEC   European Economic Community

EOS   NASA's Earth Observing System

EOSAT   Earth Observation Satellite Company (US)

ERDAS	Earth Resources Data Analysis Systems (company)
ERIM	Environmental Research Institute of Michigan
ERTS	NASA Earth Resources Technology Satellite
ESA	European Space Agency
ESRI	Environmental Systems Research Institute (US)
FAO	United Nations Food and Agriculture Organisation
GEMS	UNEP's Global Environment Monitoring System
GIS	(Computerised) Geographical Information System
GNP	Gross National Product
GSFC	Goddard Space Flight Center
GRID	UNEP's Global Resource Information Database
GTE	NASA's Global Tropospheric Experiment
IBM	International Business Machines
ICSU	International Council of Scientific Unions
INMARSAT	International Maritime Satellite Organisation (UK)
INTELSAT	International Satellite Communications Organization (US)
IRPTC	UNEP Industry and Environment Office's International Registry of Potentially Toxic Chemicals (Switzerland)
ISDN	Integrated Services Digital Network
ISIS	International Species Inventory System (Minneapolis)
ISRO	Indian Space Research Organization
IUCN	International Union for the Conservation of Nature and Natural Resources (Switzerland)
IT	Information Technology
ITU	International Telecommunications Union
ITC	International Institute for Aerospace Survey and Earth Sciences (Netherlands)
LAN	Local Area (computer) Network
LISP	List programing (computer language)
MOS-1	Japanese earth resources satellite
MSS	Multispectral (remote sensing) Scanner
NAS	National Academy of Sciences (US)
NASA	National Aeronautics and Space Administration (US)
NEC	National Electronics Corporation (Japan)
NESDIS	NOAA's National Environmental Satellite, Data, and Information Service
NOAA	National Oceanic and Atmospheric Administration (US)
NROSS	United States Navy Remote Sensing System
NTIS	National Technical Information Service (US)
OTA	Office of Technology Assessment, United States Congress
PC	Personal computer
PCC	People's Computer Company (US)
PROLOG	Programing in Logic (computer language)

RAM	Random Access (computer) Memory	TNCs	Transnational corporations
SAR	Synthetic Aperture (remote sensing) Radar	UNEP	United Nations Environment Program
SDI	United States Strategic Defense Initiative	UNICEF	United Nations Children's Fund
SHARE	INTELSAT's Satellite for Health and Rural Education	UNIDO	United Nations Industrial Development Organisation
SITE	Satellite Instructional Television Experiment (India)	US AID	United States Agency for International Development
SPOT	Systeme Probatoire d'Observation de la Terre (France)	VITA	Volunteers in Technical Assistance (US)
SRAM	Static Random Access (computer) Memory	WHAZAN	World Bank Hazard Analysis Software
TDRSS	Tracking and Data Relay Satellite System	WCED	World Commission on Environment and Development
TI	Texas Instruments Corporation (US)	WRI	World Resources Institute
TM	Thematic Mapper (remote sensor)	WRL	Willow Run Laboratories of the University of Michigan

APPENDIX: COMPANY AND ORGANIZATION PROFILES

In an enormous and rapidly growing industry, the following profiles do not aspire to complete coverage. Instead, they provide a selection of "lenses" through which to view various existing and potential applications of information and remote sensing technologies. The length of a profile does not necessarily reflect a company's or organization's relative importance. The profiles were predominantly written in early 1987.

The companies and organizations profiled represent a sample of some of the more forward-looking organizations under the following broad headings: big computer and telecommunication companies; small computer companies; national space programs; international collaborative programs; remote sensing companies; value-added companies; industry organizations; university and non-profit research organizations; and environmental data suppliers.

Institutional profiles may be found in the following order:

1. Big Computer and Telecommunications Companies
Apple
American Telephone and Telegraph Company
Control Data Corporation (US)
Cray
Digital Equipment Corporation (US)
International Business Machines

2. Small Computer Companies
People's Computer Company (US)
SAI

3. National Space Programs
National Aeronautics and Space Administration (US)
National Oceanic and Atmospheric Administration (US)

4. International Collaborative Programs
ARGOS
Communications Satellite Corporation
International Maritime Satellite Organisation (UK)
International Satellite Communications Organization (US)

5. Remote Sensing Companies
Earth Observation Satellite Company (US)

KRS
Systeme Probatoire d'Observation de la Terre (France)

6. Value Added Companies
Autometric
Earth Resources Data Analysis Systems
Environmental Systems Research Institute (US)
RDA International Inc.
Satellite Hydrology

7. Industry Organizations
American Federation of Information Processing Societies

8. University and Non-Profit Research Organizations
Cornell Laboratory for Environmental Applications of Remote Sensing
Environmental Research Institute of Michigan
Remote Sensing Research Unit (University of California)

9. Environmental Data Suppliers
Dialog
EIC/Intelligence Inc.
NOAA's National Environmental Satellite, Data, and Information Service
National Technical Information Service (US)

AMERICAN FEDERATION OF INFORMATION PROCESSING SOCIETIES (AFIPS)

Contact: John R.B. Clement, Director, Governmental Activities
Address: 1899 Preston White Drive, Reston, VA 22091
Telephone: (703) 620-8900

AFIPS is a professional association representing some 250,000 computer professionals in eleven societies. Its function is to link the computer science and engineering professions with government and to represent the computer profession in the councils of IFIP, the International Federation for Information Processing.

Although John Clement of AFIPS reports that the Federation's professional staff "have not specifically examined applications of information technology to sustainable development or environmental management, nor technology transfer to the Third World countries," AFIPS monitors key issues through its National Information Issues Panel. Such issues may relate to hardware (e.g., international negotiations about geostationary "parking orbits"), software (e.g., patent and copyright protection of the intellectual property encoded in software), or the information stored in databases (e.g., issues relating to national sovereignty, privacy and transborder data flows).

Whilst the United States champions unrestricted exchange of information, it also "increasingly restricts certain kinds of scientific and technical information deemed of strategic importance." The developing countries, meanwhile, have often criticized the "free access" and "uncensored reporting" advocated by the United States, in the belief that it leads to biased reporting, in which only the bad news is emphasized. They have called for a "World Information and Communications Order," covering both equitable access to the technology and control over, or alternatives to, foreign reporting.

AFIPS recognizes the danger that the dynamics of the emerging information economy could, in extreme cases, aggravate the gap between developed and less developed nations by dividing the world into "data poor" and "data rich" nations or regions. This widening gap could pose particular problems in the environmental field, where the most pressing needs are often in the Third World, while the relevant hardware and software are in the First and Second Worlds.

APPLE COMPUTER, INC

Chairman: John Sculley
Contacts: Paul Wiefels, Group Marketing Communications Manager (US)
Tony Fraser, Product and Marketing Manager, Apple Computer (UK) Ltd.
Addresses: Paul Weifels, 20525 Mariani Avenue, Cupertino, CA 95014 or on (408) 973-3538
Tony Fraser, Eastman Way, Hemel Hempstead, Hertfordshire HP2 7HQ, UK or on 0442 60244

Apple was started in a garage by two California drop-outs in 1976. The launch of the Apple II microcomputer in 1977 helped revolutionize the U.S. computer industry, spurring the trends toward machines with much smaller "footprints," able to sit comfortably on the desktop. But the strains of growing from a start-up company to a $2 billion-a-year corporation were tremendous. The company's first quarterly loss triggered a massive reorganization in 1985, which included laying off 20 percent of the company's employees. When John Sculley, previously with PepsiCo, ousted Apple founder Steven Jobs, the company's reputation as the "Camelot" of the computer industry seemed threatened, but Apple has continued to develop and to break into new markets.

Since the early days, Apple has always made a feature of "user-friendliness" (because it has predominantly sold into the personal computer market), an approach that characterizes its current best-seller, the Macintosh. Although Apple had only 7 percent of the U.S. corporate personal computer market by 1986, compared to IBM's 33 percent share, with the rest taken by PC "clones," Macintosh sales doubled that year. And Apple's latest ("Open Mac") Macintosh computers, built around the Motorola 32-bit microprocessor, are IBM-compatible for the first time.

A key growth area for the company has been "desktop publishing," which teams the Macintosh with the company's Laserwriter printer and "Pagemaker," a program developed by Aldus, a Californian software company. "A year ago when we started talking about desktop publishing, most people thought it would be, at best, a very small niche market," John Sculley recalled early in 1987. "Now it has exploded." In 1986 alone, Apple sold 50,000 Apple publishing systems, and it now plans to build on this success to break into other markets, including desktop communications, desktop presentation and desktop engineering. (Analysts forecast that the presentation graphics market alone will grow from $100 million in 1986 to $1 billion by 1990.)

By accepting the need for IBM-compatibility and building toward computer networking through the development of Macintosh "workgroup" networks, which can be bridged to larger, faster machines through "network gateways," Apple hopes to capture a larger share of the business computer market.

55

It is linking up with DEC to compete with IBM, but will still need to keep a close eye on the competition. William C. Lowe, President of IBM's personal computer unit, the Entry Systems Division, recently noted that IBM's future PCs will offer "a new level of user-friendliness." However, IBM's new line of PS/2 personal computers are not entirely compatible with its original PCs. Further, delays in software development for the new range are giving some corporate customers the time to think about turning to the easier-to-run Apple systems.

Interestingly, the two founders of Apple, Steven Jobs and Steven Wozniak, have both started their own companies. Jobs's Next Inc. is developing a "super workstation" personal computer aimed at the college student market; while Wozniak's Cloud Nine has already begun marketing a remote control device to operate all home electronics gear from one controller. Both enterprises have moved the founders of Apple far from the Apple vision of eleven years ago of "computer power for the masses."

ARGOS DATA COLLECTION AND LOCATION SATELLITE SYSTEM/SERVICE ARGOS, INC

President: Jean-Luc Bessis
Contacts: Archie Shaw, Executive Vice President and Director of Marketing and Promotion (US)
Philippe Courrouyan, Chief, Marketing Department (France).
Addresses: Archie Shaw, Service ARGOS Inc., 1801 McCormick Drive, Suite 10, Landover, MD 20785 or on (301) 925-4411
Philippe Courrouyan, Chief, Marketing Department, CLS Service ARGOS, 18 Avenue Edouard-Belin, 31055 Toulouse Cedex, France or on 61 27 43 51

A collaborative program between CNES, the French Space Agency, NASA, and NOAA, ARGOS is a satellite-based environmental data collection system. It locates—and gathers data from—fixed or moving platforms at any point on the world's surface, for use in meteorology, oceanography, hydrology, animal tracking, seismology, glaciology, marine biology, and so on. Continuously operational since 1979, ARGOS is expected to continue well into the 1990s.

The system, which operates 24 hours a day, 365 days a year, consists of two NOAA satellites, each equipped with an onboard data collection and location system. The recorded data are "downlinked" each time a satellite is within visibility of one of the three telemetry receiving stations (Gilmore Creek and Wallops Island in the United States; Lannion in France). The data are transmitted to the National Environmental Satellite and Data Information Service (NESDIS: see page 67) center at Suitland, Maryland. ARGOS data are separated out and transmitted on to Landover and Toulouse.

The platforms that relay data up to the satellites are equipped with sensors and a platform transmitter terminal (PTT). At any given moment, each satellite "sees" all PTTs within a 5,000 km-diameter circle beneath its path. As the satellite orbits, it sweeps a 5,000 km-wide swath covering both Poles. The swath is displaced by 25° (i.e. 2,800 km at the equator) every revolution, as a result of the Earth's rotation.

Some PTTs are located on buoys, ships, or yachts, while others are deployed by helicopter or light aircraft in the polar regions, to monitor the movement of ice floes in areas frequented by drilling rigs and shipping. The ARGOS system has also been used for animal tracking on land (e.g., bears and caribou), and at sea (e.g., whales, dolphins, seals, turtles, and basking sharks) to study species distribution and migration. The development of new, smaller, physically more robust and analytically more sensitive sensors is permitting the collection of data on animal behavior which would have been inconceivable even a few years ago. The *ARGOS Newsletter* is available free of charge, either from Service ARGOS, Inc. or CLS Service ARGOS.

AT&T (AMERICAN TELEPHONE AND TELEGRAPH COMPANY)

Chairman: James E. Olson
Contacts: A.J. Juliano, Manager, Customer Assistance; and John Leonard, District Manager
Addresses: A.J Juliano, Room 19-1E2, 222 Mt. Airy Road, Basking Ridge, NJ 07920 or on (201) 953-5285
John Leonard, Room 5356-C3, 295 North Maple Avenue, Basking

Ridge, NJ 07920 or on (201) 221-3659
Head Offices: 550 Madison Avenue,
New York, NY 10022

Following the break-up of the Bell System on January 1 1984, there emerged seven regional telephone companies, or "Baby Bells," and AT&T. The 1982 antitrust settlement with the Justice Department was supposed to open up the telecommunications industry to outside competition. Even after the break-up, however, AT&T remains a massive company, with a 1986 turnover of $34 billion, dominating the $50 billion U.S. long-distance telephone market. It has also moved into new areas, including work on such technologies as optical computing systems.

Following the Bell break-up, as AT&T Chairman James Olson explained, it soon became clear that AT&T "was on the right road but in the wrong lane." The company's basic business is information movement and management, but following a period marked by high costs and flat earnings, a decision was made to adopt a more focused strategy. This "single enterprise strategy" has three priorities: (1) to retain the company's lead in core businesses such as long distance services, communications equipment for home and office use, and network telecommunications equipment; (2) to develop computers not simply as stand-alone systems, but also as a vital element in information networks; and (3) to take the company's business into global markets, with strategic links already forged with European and Far Eastern companies. AT&T is also working with IBM to establish UNIX as the standard operating language, initially for larger computers.

Although the company took a $3.2 billion write-down in the fourth quarter of 1986, mainly the result of problems with its computer and phone equipment businesses, it announced plans to spend $2.5 billion in 1987 and up to $8 billion by 1989 on modernizing its long-distance network. Along the way to becoming competitive in a deregulated market, it has had to shed 66,000 jobs. Included in AT&T's plans are 103 giant new digital network switches, new computerized equipment to protect against network outages, and 24,000 miles of high-capacity fiber optic cable.

AT&T is developing new chips based on gallium arsenide rather than silicon under a Defense Advanced Research Projects Agency contract worth around $20 million. These chips will be faster by an order of magnitude, use less power and produce less heat than silicon chips. Production of its megabit memory chip, which can store one million bits of data, has been growing steadily. The company is also working on new information management techniques, such as "wideband packet technology," designed to dramatically expand the volume of information that can be sent down the lines—and to add data and pictures to the voice signals already transmitted.

Bell Laboratories, founded in 1925 and now within the AT&T group, are focusing on promising new concepts like lightwave technology—involving the transmission of vast amounts of information as speeding pulses of laser light. The goal is to build optical computers, which would use photons rather than electrons.

Bell Labs built the first lightwave switching chip, which actually uses light to control light. It is based on switches that are turned on and off by light beams, just as electronic transistors are turned on and off by electric charges. "This may become the main building block of a lightwave switching machine," explained optical computer researcher Alan Huang. The result would be "a 'computer' that could keep pace with speeding light pulses by switching them up to one thousand times faster than today's electronic switches." The payoff of all this work could be the ability to send 10 million conversations or 10,000 digital television channels simultaneously through a single optical fiber.

Even further along, Bell Lab researchers are working on electronic neural networks, experimental chips that may function like brain cells. "We are using models of brain function to give us new ideas on how to do computing," said Hans Peter Graf. The optical computer, if it proves possible, could run up to 1,000 times faster than today's supercomputers.

AUTOMETRIC, INC

President: Clifford W. Greve
Contact: Dr. Andrew Biache, Jr., Senior Vice President
Address: 5205 Leesburg Pike, Suite 1308/ Skyline 1, Falls Church, VA 22041
Telephone: (703) 998-7606 supercomputer

57

Founded in 1957, Autometric combines expertise in imagery interpretation, computer science, photogrammetry, and systems engineering. Its overall goal is to solve increasingly complex remote sensing problems. The company, which has reported an average compounded annual growth rate of 30 percent since 1980, is involved in military remote sensing as well as commercial applications of the technology.

Autometric has developed a number of geographical information systems, including AUTOGIS (a public domain GIS specifically designed for land use management), DELTAMAP (a proprietary turnkey GIS for industry and local governments, including users working on hazardous materials, waste disposal, and pollution monitoring), and MILGIS (a GIS designed for military and intelligence applications).

AUTOGIS has been used for environmental impact studies associated with the Central Arizona Project (CAP), a massive canal system planned to supply the Phoenix and Tucson areas with water from the Colorado River. Once the canal is in full operation, some time in the early 1990s, it will stretch 330 miles, through 14 pumping stations. The Autometric GIS project focused on the likely impacts from the construction of the New Waddell, a feeder dam just north of Phoenix. Key impacts are likely to be associated with construction activities, and subsequent flooding, water-logging, or disruption of natural drainage systems.

Bureau of Reclamation personnel have investigated potential impacts on land use, land ownership, access, recreational use, and fish and wildlife habitats. Autometric has also been investigating the use of an "expert system" as part of a GIS used to manage the Nicolet National Forest in Northern Wisconsin. By distilling the knowledge of professionals in silviculture, wildlife management, pest management, and soil science into such a system, Autometric hopes to develop a sophisticated tool for use when deciding on how to manage aspen stands for timber and/or wildlife. Autometric publishes a newsletter, *AUTOGIS*.

COMMUNICATIONS SATELLITE CORPORATION (COMSAT)

President: Marcel Joseph
Contact: Richard McGraw, Vice President of Corporate Affairs

Address: 950 L'Enfant Plaza S.W., Washington, D.C. 20024
Telephone: (202) 863-6000

COMSAT was inaugurated by the 1962 Communications Satellite Act as a for-profit company to run a global communications satellite system as part of President Kennedy's drive to master the space age. Its first satellite, Early Bird, was launched in 1965. While COMSAT concentrated mainly on serving the U.S. domestic demand for satellite communications services, it created, developed, and managed the International Satellite Communications Organization (INTELSAT) system over a period of 12 years.

INTELSAT is a consortium of 113 countries that operates a network of 16 communications satellites providing telephone, video, digital data, and TV services on a global basis. It is often described as one of the better examples of international co-operation and has been suggested as a possible model for an international earth observation organization. With a 1985 turnover of around half a billion dollars, INTELSAT has been able to introduce a number of services aimed specifically at developing countries. Project SHARE (Satellites for Health and Rural Education) and the INTELSAT Assistance and Development Fund aim to help developing countries establish telecommunication infrastructures.

The second international organization spawned by COMSAT, via its MARISAT business which dates from 1976, is the International Maritime Satellite Organization (INMARSAT), which deals exclusively with communications to and from ships, offshore rigs, and mobile earth stations (e.g. mobile telephones).

COMSAT and INTELSAT, however, have been facing fierce competition from optical fiber communication networks. Further, these pressures are likely to increase as the geostationary orbit is opened up to smaller organizations by competition among space launch companies.

INTELSAT plans to counter the optical fiber challenge with its advanced Intelsat VI satellite, due for launch in 1988 or 1989. The new satellite is expected to transmit digital data three times faster than the TAT-8 trans-Atlantic optical fiber link at one third the cost. COMSAT has met the competition on the home front by expanding into the business communication and direct-broadcast TV markets.

INTELSAT can be contacted at 3400 International Drive, N.W., Washington, D.C. 20008, (202) 944-6800; and INMARSAT at 40 Melton Street, London NW1 2EQ, England or on 01-387 9089.

CONTROL DATA CORPORATION (CDC)

President: Robert M. Price
Contact: Gerald Hendin, General Manager of External Affairs
Address: Corporate Headquarters, 8100 34th Avenue South, Minneapolis, Minnesota 55440. Mailing address: P.O. Box 0, Minneapolis, Minnesota 55440-4700
Telephone: (612) 853-8100

As Chapter VI explains, CDC has emerged from severe financial problems, shedding a considerable range of socially oriented programs. Most of them were originally launched by CDC co-founder William C. Norris, and some related to sustainable development needs (see page 43).

The company's 1985 results were the worst in its history: it lost $567.5 million, compared with a $5.1 million profit the previous year. CDC's new President, Robert M. Price, explained that its operations "simply had become too diverse. In today's environment of intense international competition and continued rapid technological change, diversification can be highly dilutive of financial resources and, even more important, management attention."

By early 1987, CDC had cut employment levels by a quarter, to 30,000, eliminating nearly half of all management jobs. As a result of such changes, the company moved back into the black in the first quarter of 1987—a trend which most analysts expected to continue. "Crisis can produce chaos," as Price told *Business Week*, "but it can also galvanize people into change." Today's CDC sees its business as providing customers with "products and services based on computer technologies," rather than pursuing Norris's goal of "serving society's unmet needs."

CDC was once *the* supercomputer company. But in the early 1970s it lost Seymour Cray, its leading supercomputer designer, who left to found Cray Research (see page 60). Since then, Cray has sold or leased nearly 150 supercomputers, accounting for around two thirds of the total market to date.

Now CDC is back in the game, through a 89 percent holding in ETA Systems, Inc. ETA shipped its first supercomputer in December 1986 and claims that its ETA-10 will run faster than Cray's top-end machines. If early teething problems with the machine are resolved, this may well prove to be the case, since the ETA-10 is designed to make 10 billion calculations a second—a speed that should cut months off some scientific calculations. ETA's automated factory, in which supercomputers are designed on computers and assembled by robots, should help cut the price of the machines.

CORNELL LABORATORY FOR ENVIRONMENTAL APPLICATIONS OF REMOTE SENSING (CLEARS)

Director: Dr. Warren Philipson
Contact: Dr. William Philpot, Acting Director
Address: CLEARS, Cornell University, 464 Hollister Hall, Ithaca, New York 14853-3501
Telephone: (607) 255-6520

The multidisciplinary nature of remote sensing work has persuaded many universities to set up special centers to assemble the various resources needed. CLEARS—a focal point for remote sensing activities at Cornell University—was established to conduct multidisciplinary research on environmental remote sensing. Administratively, CLEARS is a unit of Cornell's Center for Environmental Research.

The roots of remote sensing at Cornell date back to the 1930s, when the College of Agriculture focused on land use and land cover studies related primarily to changes in agricultural and rural economies. In the 1940s, the emphasis was on engineering landform analysis and photogrammetry. Today, however, the picture is more complicated.

"Data collection may involve optical or electrical engineering," CLEARS notes. "Data processing and analysis may involve photographic or computer science; and data interpretation may involve any discipline that makes use of the interpretations, such as agronomy, civil and environmental engineering, forestry, geology, landscape architecture, limnology, and planning. Numerous applications can be found in different disciplines and in different geographic regions. The multidisciplinary base is the principal justification for CLEARS."

Although the main focus of the CLEARS extension program is on New York State, staff research and contract work have been undertaken in most parts of the world. A free quarterly newsletter, *The CLEARS Review,* is sent to some 450 individuals and groups in approximately 45 states and 25 countries.

CRAY RESEARCH, INC

Director:	John A. Rollwagen
Contacts:	William White, Regional Support Manager (Government) (UK)
	Tina Bonetti, Director Corporate Communications (US)
Address:	William White, Cray Research (UK) Ltd, London Road, Bracknell, Berks RG12 2SY, UK, or on (0344) 485971
	Tina Bonetti, 608 Second Avenue South, Minneapolis, MN 55402 or on (612) 333-5889

Cray Research has been the industry leader in supercomputers for over a decade, though there is growing competition both from other U.S. manufacturers and from such Japanese companies as NEC. Digital Equipment (see page 61), Convex Computer, Control Data Corp's supercomputer subsidiary—ETA Systems, and other companies are cutting into the bottom end of the supercomputer market with "minisupers," minicomputers that have been upgraded so they are nearly as fast as a supercomputer but cost only about a tenth as much. IBM, too, is boosting the performance of some of its top-end mainframes to enable them to perform some supercomputer tasks. IBM is also now working with Supercomputer Systems, set up by Steve Chen, who was previously Cray's leading computer designer.[121] Today, however, about two thirds of the supercomputers installed worldwide are Cray systems—and the company's $100 million R&D budget should help keep it ahead.

The company's initial product, the CRAY-1, was first installed in 1976, and became the first commercially successful vector processor. Based on multi-

processor design, the latest Cray machines can perform multitasking, the next dimension of parallel processing beyond vectorization. The end result of these developments is that such machines can handle much more complex problems. However, Cray's self-confessed strategy is to wring the most out of existing chip technologies by innovative design. Steve Chen, top computer designer next to Seymour Cray, the company's founder, left the company in 1987 when it stopped work on its MP (multiple processor) computer. Although the MP was being designed to work at 100 times the speed of existing computers, it required fundamental research in chip design and light and laser transmission for communications within the machine. This may leave the door open to more adventurous competitors.

Many supercomputer applications have been in the military field, including projects by such overseas companies as West Germany's Messerschmitt-Boelkow-Blohm on pressure contours for a new European fighter aircraft. Cray systems have been used for geological work, including investigations of enhanced oil-recovery techniques. They have also been used to process Thematic Mapper (TM) imagery; design computer circuit boards; simulate the intricate fluid flow, heat transfer, and neutronics phenomena found in the heart of nuclear reactors; and to run the extremely complex atmospheric models needed in weather forecasting and in research on acid rain and climate change. Overwhelmingly, to date, these have been developed-world applications.

Much of the software available reflects the sort of military work that such systems are often used for. DIDSIM, for example, was developed by Sparta, Inc., to simulate incoming Intercontinental Ballistic Missiles (ICBMs) being engaged by a "Star Wars" Defense in Depth (DID) system. A considerable range of software with environmental dimensions is now available, however.

Much of this environmental software is used to model major industrial disasters, from liquefied natural gas (LNG) explosions through to dam bursts. The Argonne National Laboratory, in Illinois, has been using RADTRAN3 to calculate the radiological risk from nuclear reactor accidents in terms of early fatalities, early morbidities, latent cancer fatalities, cancer fatalities, genetic effects, and economic impacts. Scientists at NOAA's Hydrological

Research Lab have used DAMBRK to model dam failures and downstream damage. And Precision Visuals, of Boulder, Colorado, has developed CONTOURING SYSTEM, with applications not only in mining and construction, but also in environmental studies, land-use planning, and forestry.

Overseas demand for Cray supercomputers is growing. Among many other applications, some of these systems are now being used for environmental research. Britain's Natural Environment Research Council, for example, is using a Cray supercomputer to model Antarctic eddies, carbon dioxide build-up in the atmosphere, and ocean dynamics (as part of the international World Ocean Circulation Experiment). A key problem for a company such as Cray Research, however, is that a supercomputer bought by a Third World country to model monsoons, for example, could also be used to develop a nuclear weapons capability.

DIALOG INFORMATION SERVICES, INC.

Contacts: Nancy Green Maloney, Co-ordinator for Latin America; and Libby Trudell, Marketing Manager
Address: 3460 Hillview Avenue, Palo Alto, CA 94304
Telephone: (415) 858-2700

The DIALOG Information Retrieval Service, from DIALOG Information Services, Inc., has been serving users since 1972. Now, with more than 280 databases covering a very wide range of subjects, DIALOG claims to be "the most powerful online system of its type." Among the environmental databases available are ENVIROLINE (produced by EIC/Intelligence, New York, NY), ENVIRONMENTAL BIBLIOGRAPHY (Environmental Studies Institute, Santa Barbara, CA), and POLLUTION ABSTRACTS (Cambridge Scientific Abstracts, Bethesda, MD). The cost for an online search on these databases varied from $60 to $108 per connect hour in 1987.

DIGITAL EQUIPMENT CORPORATION (DEC)

President: Kenneth Olsen
Contacts: Jim Rogers, Corporate Manager of Energy and Environmental Affairs (US)

Rebecca Allen, Customer Services (UK)
Address: Jim Rogers, Corporate Headquarters, 146 Main Street, Maynard, MA 01754-2571 or on (617) 493-3837
Rebecca Allen, Digital Equipment Co. (UK) Ltd, Digital Park, Worton Grange, Imperial Way, Reading RG2 0TR, U.K. or on (0734) 868711

Founded in 1959, DEC has been the major U.S. challenger to IBM, even though its $7.6 billion turnover in 1986 was only a seventh of IBM's. But, until the 1986 merger of Sperry and Burroughs to form Unisys, DEC was the world's second largest computer company. DEC has made a remarkable recovery from its low point in 1982–83, when net income dropped from $416.2 million (1982) to $379.9 million (1983), following an ill-timed entry into the personal computer market. Interestingly, too, DEC is now linking up with Apple to compete with IBM. It is also pitching its new top-end VAX 8978, which can perform 50 million instructions per second (MIPS), against IBM's 3090-400 mainframe. DEC started out charging $4.8 million for the VAX 8978, though, just over half the price of the IBM machine.

If any one aspect of the DEC strategy distinguishes it from the pack, it is its focus on the networking of computers. "We started off by saying that we will have one computer architecture, one software system and one way of doing local area networks," Olsen recalled recently. "Each one of those decisions is dead obvious. If there's one reason why we're surviving today, it's because we did the obvious thing." That move, with all VAXes sharing the same architecture since 1978, was not as immediately obvious to IBM, which still produces machines based on a number of different architectures and operating different software systems.

What such networking means for the computer user is more computing capacity from a given number of machines. All DEC machines, from desktop machines to mini clusters of mainframe power, can be linked together in such a way that when full capacity is approached new computers can be added without a break in performance. VAXes have found many environmental applications, though mainstream applications are more likely to be found in such areas as accounting, computer-aided design

61

and manufacturing, process control, and foreign exchange dealing.

DEC is strongest in the minicomputer area, where IBM is weakest. (IBM's strength has traditionally been in the mainframe sector—where demand has fallen as customers cut back on capital spending.) Longer term, however, DEC may find its success difficult to sustain. "DEC is going through a period, much like Wang did in the 1970s and early 1980s when IBM neglected a substantial part of its market base," commented one industry analyst. "But what IBM temporarily neglects, it has a habit of reclaiming with a vengeance."

EARTH OBSERVATION SATELLITE COMPANY (EOSAT)

President: Charles P. Williams
Contact: Matthew Willard, Director, Market Planning, Applications
Address: 4300 Forbes Boulevard, Lanham, MD 20706
Telephone: (301) 552-0500

A joint partnership between RCA Corporation and Hughes Aircraft Company, EOSAT was formed to take the U.S. Landsat program into the private sector under the provisions made by the 1984 Lands Remote Sensing Commercialization Act.

With a supporting budget of $295 million to develop the next generation of Landsat satellites (Landsats 6 and 7), EOSAT is faced with the difficult task of operating the current system with revenues from the sale of remotely sensed data from the current satellites (Landsats 4 and 5). EOSAT is pinning its hopes on the value-added industry to identify a large enough market for satellite data to take the Landsat program from the development phase it is in now to full operation. Although EOSAT was allocated $250 million by the federal government to help establish a commercial market, this support has been shaken by cutbacks in government funding for the fledgling commercial space industry.

EOSAT faces strong and growing competition from the recently established French commercial remote sensing organization, SPOT Image, and from a number of government programs (in particular, the Earth Remote Sensing program of the European

Space Agency (ESA), which will become operational with the launch of the ERS-1 satellite in 1988). Japan's MOS-1 (Marine Observation Satellite) was successfully launched in February 1987, and the National Space Development Agency of Japan has plans for a more ambitious operational satellite for the 1990s. On the 25 July 1987, the Soviet Union put a remote sensing platform into space which is eight times larger than any existing remote sensing satellite. A marketing organisation, Soyuzkarta, has been set up to find markets for the images which have an even finer resolution than those from SPOT. One factor favoring EOSAT's competition is the generally higher levels of government support and funding which they receive.

Even though EOSAT is working on a novel design for the Landsat 6 and 7 satellites, allowing in-space servicing and upgrading of sensors from shuttle missions, it is plagued by federal funding uncertainties. Its hopes to capture a $50-$60 million share in the remote sensing market by 1990 may evaporate in the face of foreign competition.

Currently, the most significant private users of Landsat data are geological exploration companies, while US federal agencies account for over 60 percent of sales. The Defense Department is the biggest single customer with sales of around $10 million a year. However, EOSAT believes that the market in monitoring renewable resources, given the support of the emerging value-added industry, holds the most promise for sustained sales in satellite images. The cost barriers to Third World users remain very considerable, however, a problem which is unlikely to be eased by EOSAT's privatization.

Distribution of Landsat images is either directly through national ground stations in over 15 countries or through a network of international distribution centers. All satellite data operations are coordinated by the EROS Data Center, Sioux Falls, SD 57198 or on (800) 344 9933. *Landsat Data User Notes,* available from the headquarters in Lanham, is a newsletter reporting on EOSAT developments.

EIC/INTELLIGENCE, INC.

Contact: Terri Palmese, EIC Customer Service
Address: 48 West 38th Street, New York, NY 10018

Telephone: (212) 944-8500

Founded in 1970, as a clearinghouse for environmental isues, EIC/Intelligence has since expanded its services to cover many leading-edge technologies. Of potential interest here are online services covering artificial intelligence; biotechnology; telecommunications; energy; the environment; and acid rain.

ENVIROLINE, the company's main environmental database, is available via DIALOG (see page 61) and contains over 110,000 records, covering such areas as air pollution, chemical and biological contamination, energy, environmental education, environmental design and urban ecology, food and drugs, land use and misuse, noise pollution, non-renewable resources, oceans and estuaries, population planning and control, radiological contamination, renewable resources, solid waste, transportation, water pollution, weather modification and geophysical change, and wildlife.

EARTH RESOURCES DATA ANALYSIS SYSTEMS (ERDAS)

Contacts: Bruce Rado, Vice President; and Andrea Gernazian, Marketing Manager
Address: 430 Tenth Street, N.W., Suite N-206, Atlanta, GA 30318
Telephone: (404) 872-7327

Since 1979, ERDAS has provided integrated image processing and geographical information systems (GIS), together with consulting services in land and resource management, remote sensing, mineral exploration, and site selection. Over 400 installations worldwide use ERDAS production tested software to merge data from maps, satellites, aircraft scanners and photographs.

A family of systems is designed to meet a variety of user needs and budgets. The low-cost ERDAS-PC, based on an IBM PC/AT, operates either as a stand-alone system or as an intelligent workstation to a larger host computer. ERDAS minicomputer systems run on VAX, Prime, Data General, and Gould machines. All ERDAS software is menu-driven for easy use even by personnel without previous computer experience.

ERDAS has worked closely with government agencies and the private sector to develop image-based techniques for environmental applications. Databases have been constructed and installed for the U.S. Forest Service, U.S. Geological Survey, the United Nations, and various research institutes. ERDAS, in conjunction with ESRI, conducted a technology transfer program for the United Nations Environment Programme (UNEP) in Africa. In a collaborative effort, ERDAS and ESRI developed a link for data transfer between ERDAS and ARC/INFO systems. This has greatly increased the capability of UNEP's Global Resources Information Database (GRID) to incorporate and analyse a wide diversity of datasets.

As the developer of the first image processing GIS on a microcomputer, ERDAS continues to advocate the integration of diverse data sources as the key to global environmental decision making. Further applications of the ERDAS software are described in a newsletter, *The ERDAS Monitor*.

ENVIRONMENTAL RESEARCH INSTITUTE OF MICHIGAN (ERIM)

Contact: Dr. I.W. Ginsberg, Manager, Remote Sensing Services and Applications Laboratory
Address: P.O. Box 8618, Ann Arbor, MI 48107
Telephone: (313) 994-1200

Some 20 years ago, this private, not-for-profit, research and development organization pioneered the development of multi-spectral remote sensing for earth resources applications, and helps implement image processing and GIS systems. ERIM continues to carry out R&D work on new applications of remote sensing, and provides a wide range of education and training services. It also organizes the annual International Symposia on Remote Sensing of the Environment, as well as other events on related themes. Third World applications are frequently featured, both in the symposia and subsequent publications.

ENVIRONMENTAL SYSTEMS RESEARCH INSTITUTE (ESRI)

Director: Jack Dangermond

Contact: S.J. Camarata, Jr., Director of Marketing and Systems Coordination
Address: 380 New York Street, Redlands, CA 92373
Telephone: (714) 793-2853

Founded in 1969, ESRI has pioneered in the field of geographic information systems (GISs), and its products are now in use in over 150 locations around the world. Its ARC/INFO software has found a growing array of applications, many of them environmental. Users have included state governments and such federal agencies as the Forest Service, Environmental Protection Agency, Fish and Wildlife Service, and Soil Conservation Service. ESRI also developed a system GIS for the United Nations Environment Programme (UNEP) in 1984, which it donated at cost. The UNEP system is operated on Prime minicomputers.

The company performs four services: software design and development; software installation and support; database consulting; and database automation. The interdisciplinary nature of GIS work is highlighted by the backgrounds of ESRI's 170 staff, who are drawn from such disciplines as computer science, geography, cartography, biology, landscape architecture, business management, urban and regional planning, engineering, and forest science.

Founder Jack Dangermond has always been committed to exploring Third World applications of remote sensing and GIS technology, though Third World work accounts for only about 5 percent of ESRI's business. Among recent projects was a database design study for the Kenya Rangelands and Ecological Monitoring Unit. Originally, environmental protection and management projects accounted for 80 percent of ESRI's business, although this has now slipped to 30 percent, albeit of a much increased workload.

A major focus of ESRI's work is on making GISs user-friendly. Indeed, the hardware costs perhaps 10 percent of the GIS system cost, with database creation and other related work costing 2-3 times as much as the hardware, at a minimum. The company distributes the ERDAS Image Processing System, which can be integrated with ARC/INFO GISs, and it is also working on an IBM PC AT compatible software package which will perform the ARC/INFO data entry functions. The PC will be used to manipulate data drawn down from a larger host computer system. Applications of ESRI's systems are discussed in a newsletter, *ESRI Newsletter.*

ETA Systems, Inc. *(See Control Data)*

INMARSAT *(See Commercial Satellite Corporation)*

INTELSAT *(See Commercial Satellite Corporation)*

INTERNATIONAL BUSINESS MACHINES (IBM)

President: John F. Akers
Contacts: Christopher Bowers, Corporate Programmes, IBM (UK) Ltd
Dr. Eric Mahler, Director of Scientific Programs (US) and
Dr. P. Bai Akridge, Liaison of Government Programs (US)
Addresses: Christopher Bowers, 76 Upper Ground, London SE1 9PZ or on (01) 928-1777
Dr. Eric Mahler, Rockwood Road, Town of Mount Pleasant, North Tarrytown, NY 10591 or on (914) 332-3068
Dr. P. Bai Akridge, 1801 K Street N.W., Suite 1200, Washington, DC, 20006 or on (202) 778 5008
Corporate headquarters: Armonk, NY 10504
(See also under individual IBM Scientific Centers, below)

A profile of IBM can be found in Chapter V. In addition to its U.S. Scientific Centers in Cambridge, Massachusetts, Los Angeles, California, and Palo Alto, California, IBM has thirteen Scientific Centers outside the United States (Brazil, Egypt, France, Germany, Israel, Japan, Italy, Kuwait, Mexico, Norway, Spain, the U.K., and Venezuela), with another being considered for Norway. Their role is to "provide a gateway between the IBM Corporation and the scientific and technical community at large." Their university-like operations are aimed at developing new applications of information-processing techniques, involving close collaboration with other academic, research, and high technology institutions.

A considerable number of environment-related projects have been carried out at these Centers, from work on the modelling of major hazards (e.g. preparation of volcanic hazard maps at the Pisa Center, and investigation of consequences of liquefied natural gas (LNG) spills in the Suez Canal at the Cairo Center), through air pollution modelling (in the United States, Italy, Japan and Kuwait), to work on better ways of using Israel's water supplies (Haifa Center) and Egypt's reclaimed desert lands (Cairo Center).

For the purposes of the WRI Technology Project, three of the Scientific Centers were visited, one in the United States (Palo Alto), and two in Europe (Paris and Madrid). Work at these three centers has helped build IBM's capability in image processing and remote sensing—two areas of particular relevance to environmental management.

(1) IBM SCIENTIFIC CENTER, PALO ALTO

Contacts: Dr. Horace Flatt
              Dr. Ralph Bernstein
              Dr. Joe Myers
Address: 1530 Page Mill Road, Palo Alto, CA
              94304
Telephone: (415) 855-3261

Like the other Scientific Centers, the one in Palo Alto has an interest in broadening computer markets and operates on a minimum three-to-five year time horizon. It has worked on environmental applications of computers, though not for some years. Since environmental monitoring applications need extraordinarily fast data input rates from sensors, the Center examined the hardware and software implications.

The attraction of remote sensing to manufacturers has stemmed from its reliance on substantial computing power, particularly for image processing applications. The problem, though, as Dr. Flatt notes, is that "most of the identifiable users don't have a lot of money."

The entry costs are falling, however. Dr. Flatt pointed out that the current (late 1986) cost of a high-resolution computer system, connected with a larger host system, was around $150,000, but that this price-tag would fall to $50-$75,000 over the next few years. In conjunction with this trend, Joe Myers at the Palo Alto Center is working on ways of using IBM PCs for image processing work to pull the entry threshold down even lower.

Dr. Bernstein, well known for his work on the computer processing of satellite remote sensing data, has worked on combining different types of data, including Landsat Thematic Mapper data, geophysical gravity data, and digital line graph cartographic data. He has worked on methods for correcting the absolute and relative geometry of images; registering two or more images to each other; overlaying graphics data onto image data; changing the geometry of image and graphics data into any of 20 standard cartographic projections; and assessing the accuracy of any geometric operation.

Basic research at Palo Alto, and other European Centers has supported a number of applications in Centers in developing countries. The Kuwait Scientific Center has worked with the Kuwait Institute for Scientific Research on the use of Landsat and NOAA data on applications such as desert exploration, and detection of the Nowruz oil field spill into the Arabian Gulf.

(2) PARIS SCIENTIFIC CENTER

Contacts: J.J. Duby, Director of Science and
              Technology, IBM France
              Gerard Savary
              Dr. Diem Ho
Address: 36 Avenue Raymond Poincare,
              75116 Paris, France
Telephone: (1) 45 05 14 00

The Paris Center has been working with SPOT Image to develop software for processing satellite images, focusing on the merging of satellite and non-satellite data sets in geology and climatology. It has also been looking, in its Sahel Project, at climatic change in Africa's Sahel zone and, using Landsat data, at the consequences of such changes for vegetation cover.

The Center has produced "surface status" maps of natural resources in such countries as Burkina Faso, initially using Landsat data. These were useful at the regional scale, but attention is now on the use of SPOT data, whose definition allows for better precision. (Pixel size is 20 meters × 20 meters, versus 60 meters × 80 meters for Landsat.) The SPOT scanners can even pick up soil erosion problems, a critical consideration in natural resource management, particularly in the tropics. The image processing is

65

done on the Hacienda system, run on an IBM 7350, using software developed by IBM France.

In collaboration with the Medecins Sans Frontieres organization, the Paris Center is also working on the use of expert systems in medical decision-making and training. Such systems, which will encapsulate the experience of professionals in the field, could find important longer term applications in developing countries, to compensate for limited trained personnel.

(3) MADRID SCIENTIFIC CENTER

Contacts: Dr. Antonio Santisteban, Scientific
                Programs Manager
Address: IBM, S.A.E., Paseo de la Castellana
                4, 28046 Madrid, Spain
Telephone: (34)-1-734.21.62

Once active in processing remote sensing data, with studies carried out in such areas as the Ebro Delta, the Madrid Center has moved on to look at the applications of its techniques in processing medical images. It is working, for example, with the Centro de Biologia Molecular at CSIC (High Council for Scientific Research) on the structure and function of two viruses and with the Cardiology Department of the Reina Sofia Hospital in Cordoba on the detection of coronary stenosis. This work, like remote sensing before it, is "forcing" the technology, resulting in techniques suitable for a wide range of medical and non-medical applications.

Work has been conducted in fields other than remote sensing and image processing. The LMPS (Linear Multi-objective Programming System) developed by the Madrid Center was used in a joint project between the Egyptian Water Master Plan project and the Cairo Scientific Center in the formulation of a multi-objective model for planning the reclamation of desert lands. A small number of the 68 target sites for development were chosen as case studies to demonstrate the application of the computer techniques.

KRS REMOTE SENSING

President: Peter C. Moran
Contacts: Ms. D. Park, Executive Administrator, and Christopher K. Veronda, Corporate Communications, Eastman

Kodak Company, KO/NOD, 343
State Street, Rochester, NY 14650
or on (716) 724-5802
Address: 1200 Caraway Court, Landover, MD
                20785
Telephone: (301) 772-7800

KRS Remote Sensing is a subsidiary of the Eastman Kodak Company's Eastman Technology Inc. Set up in 1987, KRS is Kodak's frontrunner in the search for new markets being opened up by the application of remotely sensed data. Few companies of similar size and standing have targeted this market for diversification and expansion.

While the new company has been careful not to restrict its target client base to those working in renewable resources and environmental management, it does see environmental agencies wishing to assess pollution damage, and state and local governments needing imagery for land-use planning as important constituents. "Kodak wants to make remote sensing a growing market rather than take a market share" is how the parent company put it when Kodak started to put KRS together after losing the bid for the Landsat system to RCA and Hughes (see EOSAT).

The major commercial consumers of remote sensing data have, until recently, been the oil and mineral exploration companies. KRS is entering the market at a time when activity in this sector is slackening off considerably. To survive in the rapidly changing scene KRS will probably have to work hard at developing remote sensing services in previously under-developed markets. This could be good news for environmental data users, who have not always been able to pay the same sort of prices that the extraction industry has been willing to pay for its remotely sensed imagery.

KRS has signed an agreement with the Canadian company, MacDonald Dettwiler and Associates, for the use of their Meridian image processing software package. Installed on DEC computers, Meridian can combine digital imagery data from any combination of sources (e.g., Landsat, SPOT or even aircraft data) for analysis and interpretation against a wide range of specific applications. In common with other companies in the field, including ERDAS and ESRI, KRS is also looking to develop and operate customised installations for clients.

National Environmental Data Referral Service (NEDRES) *(See National Environmental Satellite, Data, and Information Service.)*

NATIONAL AERONAUTICS AND SPACE ADMINISTRATION (NASA)

Administrator: James C. Fletcher, Ph.D.

Contacts: Dr. Robert Watson, Global Habitability Program, EE NASA Headquarters, Washington, DC 20546 or on (202) 453-1681

Mr. James T. Rose, Assistant Administrator, NASA Headquarters, Office of Commercial Programs, Washington, DC 20546 or on (202) 453-1123

Mr. Ross Nelson, Code 623, NASA Goddard Space Center, Earth Resources Branch, Greenbelt, MD 20771 or on (301) 286-9925

Dr. Lennard A. Fisk, Associate Administrator, Office of Space Science and Applications, NASA Headquarters, Washington D.C. 20546 or on (202) 453-1409

NASA often talks in terms of "spinoff" benefits from its programs. An estimated 30,000 secondary applications of aerospace technology have emerged in the nearly 30 years since NASA was founded. Through its Congressionally-mandated Technology Utilization Program, NASA promotes the broader and accelerated use of its ever-growing bank of technical knowledge, publishing a directory called *Spinoff.*

In the environmental field, NASA expertise and technology have helped on many fronts, from getting earth observation systems such as the Landsat and Seasat satellites into orbit, through work with the U.S. Agency for International Development (AID) on the use of Landsat data to map trends in urban encroachment onto Egyptian farmland, to research on cutting spray drift from aerial crop spraying. Computer models developed by NASA have also found other applications, such as those supplied by the Computer Software Management and Information Center (COSMIC) for use in the design of steam power plants in India.

Although spinoffs are clearly important, NASA's longer-term contribution to environmental protection and sustainable development may well come through its mainstream activities—including the successors to NASA's "Global Habitability Program," first proposed in 1982 at UNISPACE-82. As NASA sees them, the basic aims of such research are to: (1) understand the vital global processes of the earth's energy balance, the global hydrological cycle, and the biogeochemical cycling of carbon, nitrogen, phosphorus, and sulfur; (2) obtain quantitative measurements of long-term changes in these processes; and (3) assess the impact of such long-term changes on the continued habitability of the earth by man and other species.

The basic idea is now being put forward in various guises—as "Earth Systems Science" at NASA, "Global Geosciences" at the National Science Foundation (NSF), and as "International Geosphere-Biosphere Program" at the National Academy of Sciences. But as *Science* explained (5 September 1986, p. 1040), "each name expresses the same fundamental idea: a simultaneous study of the climate, the oceans, the biosphere, the dynamics of the solid earth, and the biogeochemical cycles of all the major nutrients—in short, a study of the earth as an integrated whole."

To study processes that unfold over decades or centuries, like the atmospheric build-up of "greenhouse" gases, such programs will need a permanent network of satellites, coupled with another network of instruments on the ground, all feeding data into state-of-the-art computers. According to NASA's Office of Space Science and Applications, NASA's portion of the global space research program could be accommodated within an earth observations budget some 25 percent higher than the 1986 level of $300-$400 million.

NATIONAL ENVIRONMENTAL SATELLITE, DATA, AND INFORMATION SERVICE (NESDIS)

Contact: Dr. Gregory W. Withee, Acting Director, Information Service Center or Russ Koffler (202) 763-7190 [Dr. Joan Hock (formerly Director of the NESDIS Information Service Center), Chief, Technological Hazards Division, Federal Emergency Management Agency, 500 C Street,

S.W., Room 630, Washington, DC 20472 or on (202) 646-2860]

Address: NESDIS, Universal Building South, 1825 Connecticut Avenue N.W., Room 511-B, Washington DC, 20235

Telephone: (202) 673-5394

The National Oceanic and Atmospheric Administration (NOAA) gathers worldwide environmental observational data about the oceans, earth, air, space, and sun and their interrelationships. Once these data have served their purpose for NOAA or some other collecting agency, they are passed on to four service centers maintained by NOAA as part of NESDIS.

The four centers are the Assessment and Information Services Center (AISC) at the address given above; the National Climate Data Center (NCDC), Federal Building, Ashville, N.C. 28801 or (704) 259-0682; the National Geophysical Data Center (NGDC), 3100 Marine Street, Boulder, CO 80303 or (303) 497-6215; and the National Oceanographic Data Center (NODC), Page Building 1, Room 428, 2001 Wisconsin Avenue, N.W., Washington, D.C. 20235 or (202) 634-7500.

NOAA's National Environmental Data Referral Service (NEDRES), which has been developed since 1980, provides an access point to all these environmental data. Operated by AISC, it is a publicly available computerized catalog and index that identifies the existence, location, characteristics, and availability of environmental data. It is operated by the AISC.

NATIONAL OCEANIC AND ATMOSPHERIC ADMINISTRATION (NOAA)

Administrator: Anthony J. Calio
Contact: Dr. Gregory W. Withee, Acting Director, Information Service Center or Russ Koffler (202) 763-7190 [Dr. Joan Hock (formerly Director of the NESDIS Information Service Center), Chief, Technological Hazards Division, Federal Emergency Management Agency, 500 C Street, S.W., Room 630, Washington, DC or on (202) 646-2860

Address: Assessment and Information Services Center, 3300 Whitehaven Street, N.W., Washington, DC 20235

Telephone: (202) 634-7251

NOAA gathers worldwide environmental data about the oceans, earth, air, space, and sun and their interactions to describe and predict the state of the physical environment. It maintains four service centers as part of the National Environmental Satellite, Data, and Information Service (NESDIS: page 67), which collectively hold one of the most important U.S. environmental databases.

NOAA data are fed into many development projects in the Third World, with much of NOAA's current capability resulting from the 1973 LACIE program (Large Area Crop Inventory Experiment) and the later AgRISTARS (Agriculture and Resource Inventory Through Aerospace Remote Sensing). NOAA works with the U.S. Agency for International Development (AID) to prepare country agroclimatic assessments.

"Expanded capabilities of microcomputers have enhanced training in satellite applications in the developing world," said Dr. Hock. Personal computers are used to display assessments based on remote sensing imagery, with a growing focus on the Sahel region of Africa and on Southeast Asia.

NOAA, however, has been in a weak political position, as a subagency within the Department of Commerce. The White House Office of Management and Budget has constantly battled with Congress to cut back NOAA's operations to one orbiting weather satellite rather than two, though each year Congress votes the axed programs back into NOAA's budget.

With these pressures aggravated by the Gramm-Rudman-Hollings budget initiative, designed to cut the national deficit, NOAA has welcomed the growing emphasis on global environment research. (See NASA profile.) "When you think of it," as William Bishop, former Head of NOAA's Satellite and Information Service put it to *Science* (5 September 1986, p. 1042), "NOAA is essentially an information agency. We collect the global data set on the environment. So far, that data has been mostly for purposes of forecast and warning. But the scientific community is just as legitimate a customer for that routine data as the TV stations."

The problem, he noted, is that "a scientist has to be able to compare data over a decade or more to look for trends, which puts demands on calibration, record-keeping, and continuity that just aren't there when you focus on day-to-day prediction. It's something that we at NOAA have had difficulty stepping up to. And yet the message we get from all over Washington is that this is a *very* good charter for NOAA."

To open up its vast databases to outside researchers, NOAA plans two initiatives. NOAAPORT will provide access to the real-time data used for forecasting, while NOAANET will provide access to NOAA's long-term archival data, much of which will be made available in digital form for immediate access.

NTIS (NATIONAL TECHNICAL INFORMATION SERVICE)

Director: Dr. Joseph F. Caponio
Contacts: Dr. Joseph F. Caponio, Director (US); and Nicola Fish, Marketing (UK)
Addresses: Dr. Joseph F. Caponio, United States Department of Commerce, 5285 Port Royal Road, Springfield, VA 22161 or on (703) 487-4600
Nicola Fish, Microinfo Ltd., P.O. Box 3, Alton, Hampshire, UK or on (0420) 86848

NTIS, an agency of the U.S. Department of Commerce, is the central source for the sale of U.S. Government-sponsored research, development, and engineering reports. It is also the central source for federally generated machine-processable data files and software. It operates the Federal Software Exchange Center, which ensures the exchange of computer software between federal agencies. (The public has access through NTIS.)

NTIS publishes a range of newsletters of environmental interest. *Environmental Pollution & Control* covers air, noise, solid wastes, water, pesticides, radiation, environmental health and safety, and environmental impact statements. NTIS also publishes computer-generated bibliographies covering a wide range of environmental subjects. A catalog of published searches is available, from "Acid Rain" through to "Zooplankton." The service as a whole, however, is under considerable budgetary pressure, following public expenditure cut-backs.

PCC, INC.

Executive Director: Jane Nissen Laidley
Address: 2682 Bishop Drive, Suite 107, San Ramon, CA 94583
Telephone: (415) 833-8604

The People's Computer Company (PCC) was formed in 1972 to study the impact of the then new microcomputer technology and to act as an international educational resource. Along with the *Whole Earth Catalog,* PCC was an off-shoot of the Portola Institute. A non-profit organization, it focuses on public access to computers and computer literacy projects. PCC has worked on educational and training programs for Apple, IBM, Hewlett-Packard, Atari, Texas Instruments, and other companies.

In the California State Library Adult Microcomputer Library Project, initiated in 1983, PCC designed curricular materials and provided on-site training for librarians and library staff, including training in word processing, databases, spreadsheets, online database use, and desktop publishing. Next in line is a program on CD-ROM technology, with recent developments suggesting that it is only a matter of time before it is possible to marry text, graphics, and moving pictures in a CD-ROM database.

PCC also developed training materials for Apple Community Affairs, helping provide the basis for the Apple's Corporate Grants Program for non-profit organizations. In the COMPUTERTOWN, USA project, PCC developed public access programs for an international network of access sites, such as libraries and museums.

Although PCC has not carried out projects in the sustainable development field, Jane Nissen Laidley points out that "the issues that provoked the beginning of PCC are the very issues the World Resources Institute is addressing in its Technology Project, focusing on the implications of new technology, particularly for the Third World."

RDA INTERNATIONAL, INC.

President: Kenneth B. Craib
Address: 801 Morey Drive, Placerville, CA 95667
Telephone: (916) 622-8841

Originally established in 1972 to provide technical assistance and consulting services in resource assessment and management to developing countries, the company changed its name from Resource Development Associates to RDA International, Inc., in 1986. Like others working in the private sector side of remote sensing, Ken Craib had worked with NASA. RDA has worked in over 50 countries for a wide range of clients in the environmental and natural resource areas, including the Environmental Protection Agency, the Agency for International Aid, the Department of Defense, the Office of Technology Assessment, foreign governments, and private firms.

Of particular interest to RDA has been the implications of satellite remote sensing data for the relationships between transnational corporations (TNCs) and developing nations. Although Matthew Willard and Ken Craib observed in 1982 in a report that although it was uncertain whether remote sensing data was tipping the balance of advantage in favor of TNCs in international resource negotiations, the conclusion was that developing nations had used—and should use—satellite data to deal with TNCs on more equal terms.

Since the Law of the Sea Treaty led to most coastal states adopting 200-mile Exclusive Economic Zones (EEZs), RDA has been helping a growing number of developing countries assess and manage their fisheries. Much of the work has been carried out for AID. The main information needs usually relate to the size of the fish stocks, but RDA also lays stress on the need to determine the *sustainable* yields for such fisheries.

Although satellite remote sensing is no panacea for marine resource management, multi-spectral scanners can identify upwelling areas. Their strong thermal and spectral gradients, caused by the colder upwelling water and its different nutrient and chlorophyll content, can be traced. Overall, however, RDA concludes that fisheries applications of remote sensing data are still at a highly experimental stage, so much of the work the firm does for developing countries uses more traditional survey and analytical techniques.

REMOTE SENSING RESEARCH UNIT, UNIVERSITY OF CALIFORNIA

Director: Dr. John E. Estes, Professor of Geography

Contact: Dr. Jeffrey L. Star
Address: Department of Geography, University of California, Santa Barbara, CA 93106
Telephone: (805) 961-3845

Administratively, RSRU is a unit of the Geography Department, in the College of Letters and Science. RSRU focuses its efforts on research and training in the uses of high technology (including remote sensing, image processing and analysis, geographic information systems, and artificial intelligence) for renewable resource analysis and management. Collaborators include members of the Biology, Computer Science, Economics, Environmental Studies, and Geology departments.

Recent projects have included the development of a regional geographic database in northern Italy, information systems research for the National Aeronautics and Space Administration, earthquake hazard assessment for the state of California, development of land use management strategy for the U.S. Marine Corps, and training classes for the Environmental Protection Agency.

SATELLITE HYDROLOGY INC (SHI)

Directors: Donald R. Wiesnet and Morris Deutsch
Contact: Donald R. Weisnet
Address: 103 Beulah Road N.E., Vienna, VA 22180
Telephone: (703) 281-0216

One way to make a mark in remote sensing is to focus on a particular application area. SHI defines its mission as solving water-resources, coastal-zone, and environmental problems, but water is the key to most of the work its principals do. Wiesnet and Deutsch both gained considerable experience with the U.S. Government before branching out on their own, and both were Principal Investigators for NASA in various earth observation aircraft and satellite programs.

"Satellites, computers, and the constantly evolving technology of remote sensing have provided powerful new tools for observing the hydrosphere and man's impact upon it," they note. "Further, satellite technology now permits many traditional types of water-resource and environmental investigations and monitoring to be performed more rapidly, accurately, and effectively than ever before—with great confidence and at less cost." Administrators, resource managers, and others faced with uncertain water supplies, floods, and droughts certainly can benefit from the technology, but there remain many potential applications that the technology's cost drives out of reach.

SPOT IMAGE CORPORATION

President: Pierre Bescond
Contacts: Gerard Brachet, President, SPOT Image (France)
David S. Julyan, Vice President of Sales and Marketing, Spot Image Corporation (US)
Addresses: Gerard Brachet, 18 Avenue Belin, 31055 Toulouse Cedex, FRANCE or on (61) 424-7256
David S. Julyan, 1897 Preston White Drive, Reston, VA 22091-4326 or on (703) 620-2200

SPOT Image was incorporated in July 1982 as the first company to commercially distribute earth resources remote sensing data. Set up as a consortium of French, Belgian, and Swedish space companies and government agencies, SPOT has the task of developing a commercial market for remotely sensed data from the Systeme Probatoire d'Observation de la Terre (SPOT) satellite, launched by the French National Space Center (CNES) in 1986.

With this kind of backing, SPOT Image has been able to enter the marketplace with products that have a much greater ground resolution than those of any civilian satellite. The SPOT satellite can also be programmed to minimise cloud cover and to provide stereoscopic images. Although the first SPOT satellite uses sensors manufactured by Fairchild of the United States, the next generation will be made by France's Thomson group. SPOT has been able to make better use of the technology because, unlike U.S. civilian satellites, it is unconstrained by the Pentagon restrictions on U.S. civilian satellites not to register anything smaller than 100 feet. Interestingly, the U.S. Defense Department is one of its U.S. clients.

SPOT Image, Inc., is the U.S. subsidiary and the sole U.S. distributor for SPOT data. It competes directly with EOSAT, the commercial arm of the U.S. Landsat program. With the next generation of SPOT satellites already developed, the company is well placed to offer an uninterrupted service. EOSAT, due to launch delays on Landsat 6, is not.

With sharper images than Landsat, SPOT has been able to move into markets that have not been tapped commercially before. These include supplying media and national defense department markets. SPOT images, for example, have illustrated newspaper reports on the Chernobyl disaster and the Iran/Iraq war. In its first sales year, the company sold over 10,000 satellite pictures, and expects to become profitable by the mid-1990s.

Nevertheless, to ensure that SPOT is well placed in what is seen as a $2-$4 billion market for remotely sensed data by the turn of the century, the company is, like EOSAT, relying heavily on the value-added industry to find applications for its products in geological exploration, renewable resources, and coastal zone management. Although Gerard Brachet does not see a steep growth curve for remotely sensed data, he says SPOT is making significant efforts to broaden its applications base, particularly through collaborative programs in developing countries. SPOT Image Corporation publishes a quarterly newsletter, *Spotlight*.

SYSTEMS APPLICATIONS, INC (SAI)

President: Dr. Mihajlo Mesarovic
Address: Suite 305, Chagrin Plaza East, 23811 Chagrin Boulevard, Beachwood, OH 44122
Telephone: (216) 831-0366

SAI is a software development and consulting company that specializes in modeling and decision support systems. It was founded by Dr Mesarovic in 1971, and has an associated Canadian company, Educational Systems Applications, Inc. (ESAI) in Toronto, Ontario.

The first modeling software package (Mesarovic/Pestel model) developed by SAI was a cooperative exercise with the Case Western Reserve University and the Technical University of Hanover in West Germany. One of the first large scale global models, it represented the world in terms of ten interelated regions, and was used to prepare the second report to the "Club of Rome," *Mankind at the Turning Point.*

This model was then refined, specifically for use by government and international agencies to deal with the policy development, and then a third generation model (FORESIGHT/WIM) was designed for global modeling in business and industry. The business model has been applied, for example, to a multinational corporation with three divisions, located in the U.S., West Germany and Brazil, to evaluate alternative strategies for sustained growth. Other applications have included a study of the Lake Erie catchment area to develop strategies for the control of phosphorous pollution, and to provide national development models for Brazil, India, Mexico and West Germany.

ESAI in Canada has refined the global model for educational use. This version, called "Quest for Harmony," although based on the same representation of global conditions as the original model, can run on microcomputers. Pupils are readily able to investigate the inter-relationships between global regions, and how decisions on issues such as deforestation, population growth, and energy use can effect the global economic system. It has also been adapted for use in senior management seminars and training courses in the international business environment.

REFERENCES

1. The World Commission on Environment and Development, *Our Common Future,* (Oxford UK: Oxford University Press, 1987).

2. World Resources Institute, *The Global Possible: The Statement and Action Agenda of an International Conference on Resources, Development, and the New Century* (Washington, D.C.: World Resources Institute, 1984).

3. John Elkington, *Double Dividends? U.S. Biotechnology and Third World Development* (Washington, D.C.: WRI Paper 2, World Resources Institute, 1986).

4. Congress of the United States, Office of Technology Assessment, *Information Technology R&D: Critical Trends and Issues (Summary)* (Washington, D.C.: February 1985).

5. Ben R. Finney, "Space Development or Space Disruption? High Technology and the Third World," *Earth-Oriented Applications of Space Technology,* 1983, vol. 3, no. 2: 103–107.

6. John Lamb, "Computers Close the Gap Between Rich and Poor," *New Scientist,* September 18, 1987: 37.

7. "Information Technology: A South Special Report," *South* Magazine, December 1986, vol. 74: 51–60.

8. "The Economy of the 1990s: Seven Wary Views from the Top," *Fortune,* February 2, 1987: 48–53.

9. "Information Technology," *The Economist,* July 12, 1986: 5–28.

10. "Information Technology: Unshackling European Companies," *International Management,* February 1987: 23.

11. Allen A. Boraiko, "The Chip: Electronic Mini-Marvel that is Changing Your Life," *National Geographic,* Vol. 162, No. 4, October 1984: 421–476.

12. Bro Uttal, "How Chipmakers Can Survive," *Fortune,* April 13, 1987: 57–58.

13. Steve Dickman, "Giving Computers an Elephant's Memory," *Business Week,* September 1, 1986: 66.

14. "Zoom! Here Come the New Micros," *Business Week,* December 1, 1986: 72–78.

15. Frank Brown, "IBM Spots the Way to Smaller Chips," *The Times,* September 16, 1986: 21.

16. John Newell, "University Team Builds Bargain Supercomputer," *The Times,* February 26, 1987: 14.

17. David Hebditch and Nick Anning, "Computers Paint a New World," *New Scientist,* January 22, 1987: 33–36.

18. "Computer Industry in U.S. Still Struggling," *Wall Street Journal (Europe),* January 28, 1987: 8.

19. William M. Bulkeley and Brenton R. Schlender, "Digital, Apple to announce products to link computers," *Wall Street Journal*, January 13, 1988: 7).

20. John W. Wilson, "Suddenly the Heavyweights Smell Money in Computer Networks," *Business Week*, April 27, 1987: 82–83.

21. Louise Kehoe, "Manufacturers Struggle to Adjust," *Financial Times*, December 1, 1986: World Telecommunications Supplement.

22. "The Rewiring of America," *Business Week*, September 15, 1986: 102–107.

23. Robert W. Lucky, "Message by Light Wave," *Science 85*, November 1985: 112–113.

24. "And Now, the Age of Light," *Time*, October 6, 1986: 34–35.

25. Anthony B. Tucker, "Military and Aerospace Technology," *High Technology*, July 1986: 57–59.

26. David Fishlock, "Bell Labs Concentrates Its Light on Pushing Back the Frontiers of Telecommunications," *Financial Times*, February 12, 1987: 12.

27. Jeffrey Bairstow, "CD-ROM: Mass Storage for the Mass Market," *High Technology*, October 1986: 44–51.

28. Philip Elmer-DeWitt, "From Mozart to Megabytes," *Time*, March 16, 1987: 50.

29. "Technology Brief—Will Optical Discs Ever Forget?," *Economist*, September 26, 1987: 80–81.

30. Frost & Sullivan, Inc. "Optical Disk Drives and the Media," market study, New York, 1987.

31. Emily T. Smith and Larry Armstrong, "An Optical Memory that Can Be Wiped Clean," *Business Week*, June 15, 1987: 48–49.

32. W.W. Hutchison, "Spatially Distributed Data—A Geodetic Framework and Electronic Atlas," in *The Role of Data in Scientific Progress*, P.S. Glaeser (ed.) (North-Holland: Elsevier Science Publishers B.V., 1985).

33. R.B. King, *Remote Sensing Manual of Tanzania*, (Surbiton, Surrey, U.K.: Land Resources Development Centre of the UK Overseas Development Administration, 1984).

34. International Resource Development, Inc., "Fiber Optic Sensors," market survey report, Norwalk, Connecticut, 1986.

35. Stephen Kreider Yoder, "Japan Leads in Marketing of Bioelectronic Products," *Wall Street Journal (Europe)*, December 23, 1986: 7.

36. H. Joseph Myers and Ralph Bernstein, "Image Processing on the IBM Personal Computer," *Proceedings of the IEEE*, June 1985, vol. 73, no. 6.

37. Congress of the United States, Office of Technology Assessment, *Information Technology R&D: Critical Trends and Issues (Summary)* (Washington: February 1985).

38. Dwight B. Davis, "Artificial Intelligence Enters the Mainstream," *High Technology*, July 1986: 16.

39. William M. Bulkeley, "Technology: Expert Systems," *Wall Street Journal (Europe)*, December 9, 1986: 1.

40. M. Mitchell Waldrop, "Artificial Intelligence Moves into Mainstream," *Science*, July 31, 1987: Vol. 237, 484–485.

41. "They're Here: Computers that 'Think,'" *Business Week*, January 26, 1987: 63–64.

42. Jeff Hecht, "Computing with Light," *New Scientist*, October 1, 1987: 45–48.

43. "US Superconductor Tests Confirm Use for Computers," *Wall Street Journal*, 5 October 1987: 10.

44. Nancy E. Pfund, "The Quiet Revolution: Analytical Instrumentation Extends its Reach," research report, Hambrecht & Quist, Inc., San Francisco, April 1986).

45. Alan S. Miller and Irving M. Mintzer, *The Sky is the Limit: Strategies for Protecting the Ozone Layer* (Washington, D.C.: World Resources Institute, 1986).

46. Lisa Martineau, "Chips are Down Over Electronic Pollution," *Guardian* (UK), May 5, 1987: 8.

47. John E. Estes and Jeffrey L. Star, "Support for Global Science: Remote Sensing's Challenge," University of California: Geocarto International 1, 1986.

48. David Baker, "Remote Future for Third World Satellite Data," *New Scientist,* October 22, 1987: 48–51.

49. John W. Anderson, "Remote Sensing Finds Down-to-Earth Applications," *Commercial Space,* Fall 1985: 70–76.

50. Barry Rosenberg, "Oceanographic Images," *Commercial Space,* Spring 1986: 42–45.

51. C.J. Tucker, J.R.G. Townsend, and T.E. Goff, "African Land-Cover Classification Using Satellite Data," *Science,* January 25, 1985, vol. 227: 369–375, and R. Nelson and B. Holben, "Identifying Deforestation in Brazil Using Multiresolution Satellite Data," *International Journal of Remote Sensing,* 1986, vol. 7, no. 3: 429–448, and C.J. Tucker, J.A. Gatlin, and S.R. Schneider, "Monitoring Vegetation in the Nile Delta with NOAA-6 and NOAA-7 AVHRR Imagery," *Photogrammetric Engineering and Remote Sensing,* January 1, 1984, vol. 50, no. 1: 53–61, and H. Yates, A. Strong, D. McGinnis Jn, and D. Tarpley, "Terrestrial Observations from NOAA Operational Satellites," *Science,* January 31, 1986, vol. 231: 463–470.

52. "FAO Precautions Fail to Prevent Locust Swarms," *New Scientist,* 19 November 1987: 28.

53. United Nations Food and Agriculture Organisation, *Remote sensing applied to renewable resources* (Rome: FAO Remote Sensing Center, 1982).

54. United Nations Food and Agriculture Organisation, *Tropical Forest Resource Assessment Project: Forest Resources of Tropical Africa* (Rome: FAO, 1981).

55. H.A. van Ingen Schenau, R.J. Nicolai, J.C. Venema, F.L. van der Laan and M. Versteeg, *System Definition of Africa Real Time Environmental Monitoring Using Imaging Satellites - The ARTEMIS System,* (Netherlands: Ministry of Foreign Affairs/International Development Corporation of the Netherlands Government, for the U.N. Food and Agriculture Organisation, 1986).

56. Personal communication from Dr. M. Gwynne, Director, Global Environment Monitoring System, United Nations Environment Programme, Nairobi, Kenya, September 1986.

57. M.E. DeVries, "Use of GIS to Integrate Remote Sensing and Other Natural Resource Data," paper presented at the *Nineteenth International Symposium on Remote Sensing of Environment,* (Ann Arbor, Michigan: October 21–25, 1985).

58. Praful D. Bhavsar, "Indian Remote Sensing Satellite-utilization Plan," *International Journal of Remote Sensing,* Vol. 6, No. 3&4, 1985: 591–597.

59. Organisation for Economic Co-operation and Development, *Information and Natural Resources,* Paris 1986.

60. National Academy of Sciences, *Report of the Research Briefing Panel on Remote Sensing* (Washington, D.C.: 1985).

61. Jonathan B. Tucker, "Military and Aerospace Technology," *High Technology,* July 1986: 57–59.

62. John E. Estes and Jeffrey L. Star, "Support for Global Science: Remote Sensing's Challenge," University of California, Geocarto International 1, 1986.

63. Dennis Kneale, "Into the Void: What becomes of data sent back from space? Not a lot as a rule," *Wall Street Journal,* January 13, 1988: 1.

64. W.R. Philipson, "Problem-Solving with Remote Sensing: an Update," *Photogrammetric Engineering and Remote Sensing,* January 1986, vol. 52, no. 1: 109–110.

65. Earth Observation Satellite Company, *Directory of Landsat-Related Products and Services: International Edition* (Washington: EOSAT, 1987).

66. Daniel Charles, "US draws a veil over 'Open Skies,'" *New Scientist,* 5 November 1987: 28.

67. John H. McElroy, Jennifer Clapp, and Joan C. Hock, "Earth Observations: Technology, Economics and International Cooperation," paper presented to the National Academy of Engineering and Resources for the Future symposium on "Explorations in Space Policy," Washington, D.C., 24–25 June 1986.

68. William Ambrose, "Telecommunications: The Multiplier Effect," *Newsweek*, Telecommunications Supplement, October 19, 1987.

69. Sean Eamon Lalor, *Overview of the Microelectronics Industry in Selected Developing Countries*, United Nations Industrial Development Organization, limited distribution paper, UNIDO/IS.500, December 13, 1984.

70. Mikael Stern, "Census from Heaven? Population Estimates with Remote Sensing Techniques," in Lennart Olsson, ed., *An Integrated Study of Desertification—Applications of Remote Sensing, GIS and Spatial Models in Semi-Arid Sudan*, University of Lund, Sweden, 1985.

71. "Holding Back the Private Sector," *Economist*, October 31, 1987: Thailand Survey Supplement, p9.

72. Allan Falconer and Victor A.O. Odenyo, "Responses to Satellite Remote Sensing Opportunities in East and Southern Africa," *Advanced Space Research*, vol. 4, no. 11, 1984: 19–29, and personal communication, April 1987.

73. E.K. Wahome, *Soil Erosion Types and their Distribution in Machakos District*, Technical Report No. 126, Kenya Rangeland Ecological Monitoring Unit, Nairobi, August 1986.

74. Matthew Willard and Kenneth Craib, *Observations on Transnational Corporations, Developing Countries and Remote Sensing Data*, (Diamond Springs, California: Resource Development Associates, October 1982).

75. Anil K. Jain, *Expert Systems: Prospects for Developing Countries*, United Nations Industrial Development Organization, IPCT.41(SPEC), September 23, 1987, and Robert N. Coulson, L. Joseph Folse, and Douglas K. Loh, "Artificial Intelligence and Natural Resource Management," *Science*, Vol. 237, 17 July 1987, pp. 262–267.

76. R.B. Martin, J.R. Caldwell, and J.G. Barzdo, *African Elephants, CITES, and the Ivory Trade* (Lausanne, Switzerland: Secretariat of the Convention on International Trade in Endangered Species of Wild Fauna and Flora, 1986).

77. Mohan Munasinghe, "Practical Application of Integrated National Energy Planning (INEP) Using Microcomputers," *Natural Resources Forum*, vol. 10, no. 1, February 1986: 17–38.

78. Irving M. Mintzer, *A Matter of Degrees: The Potential for Controlling the Greenhouse Effect* (Washington D.C.: World Resources Institute, Research Report 5, April 1987).

79. G. Foley and P. Moss, *Improved Cooking Stoves in Developing Countries* (London: Intermediate Technology Development Group & Earthscan, 1983).

80. W.C. Clark and R.E. Munn (eds.), *Sustainable Development of the Biosphere* (Cambridge, UK: Cambridge University Press, 1987).

81. Brian O'Reilly, "Apple Finally Invades the Office," *Fortune*, November 9, 1987: 36–39.

82. Christopher Flavin, *Electricity for a Developing World: New Directions* (Washington D.C.: Worldwatch Institute, Worldwatch Paper 70, June 1986).

83. T.E. Beaumont, "Interpretation of Landsat Satellite Imagery for Regional Planning Studies," in *Remote Sensing in Civil Engineering*, Kerrie and Mathews (eds.) (Halsted Press, 1986).

84. Carol J. Loomis, "IBM's Big Blues: A Legend Tries to Remake Itself," *Fortune*, January 19, 1987: 34–41.

85. Frost & Sullivan, Inc., "Ten Year Strategic Analysis of IBM," market survey report, New York, 1986.

86. "IBM: Trying to Put All the Pieces Together," *Business Week*, April 21, 1986: 96–97.

87. David Mercer, *IBM: How the World's Most Successful Corporation is Managed* (London: Kogan Page, 1986).

88. Richard Thomas DeLamarter, *Big Blue: IBM's Abuse of Power* (London: MacMillan, 1987).

89. "Computing & Conservation for Sustainable Development," *IBM UK News,* Special Supplement, Issue 349, November 7, 1986.

90. World Wildlife Fund UK, *Conservation and Development Programme for the UK* (London: Kogan Page Ltd, 1983).

91. Brian Johnson, *IBM (UK) Ltd: A Case Study in Sustainable Development* (London: Centre for Economic and Environmental Development, 1986).

92. International Business Machines, *Environmental Protection,* IBM Corporate Policy Number 129A, 24 May 1973, replacing Policy 129 of 26 May 1971.

93. Guy de Jonquieres, "Inside IBM: Seeking Salvation from Software," *Financial Times,* February 2, 1987: 11.

94. Guy de Jonquieres, "Big Blue Ready to Hunt with the Pack," *Financial Times,* January 19, 1987: 14.

95. H.P. Flatt, "Computer Modeling in Energy and the Environment," *IBM Journal of Research and Development,* vol. 25, no. 5, September 1981: 571–580.

96. H. Joseph Myers and Ralph Bernstein, "Image Processing on the IBM Personal Computer," *Proceedings of the IEEE,* vol. 73, no. 6, June 1985: 1064–1070.

97. International Business Machines Corporation, *The Guide to Software for Developing Countries: A Resource for Users of IBM Personal Computers* (Neuilly sur Seine, France: IBM, 1985). Inquiries to: Communications and External Programs Manager, IBM Area South, 190 Avenue Charles de Gaulle, 92523 Neuilly sur Seine, France.

98. IBM Europe SA, *IBM in Africa* (22 route de la Demi-Lune, 92800 Puteaux, Hauts-de-Seine, France: IBM, March 1987).

99. Sam Cole, "The Global Impact of Information Technology," *World Development,* 1986, vol. 14, no. 10/11: 1277–1292.

100. Alan Cane, "The Wonder of the Software World," *Financial Times,* March 19, 1987: 37.

101. John Elkington, *Double Dividends? U.S. Biotechnology and Third World Development* (Washington, D.C.: WRI Paper 2, World Resources Institute, 1986).

102. Control Data Corporation, "Control Data: Addressing Society's Major Unmet Needs as Profitable Business Opportunities," Company policy paper, Minneapolis, 1982.

103. Patrick Houston, "How Bob Price is Reprogramming Control Data," *Business Week,* February 16, 1987: 62–63.

104. Terry Dodsworth, "Control Data Signs Eight-Year Accord with India," *Financial Times,* March 11, 1987: 6.

105. "Information Technology: A South Special Report," *South* Magazine, December 1986, vol. 74: 51–70.

106. John Lamb, "Computers Close the Gap Between Rich and Poor," *New Scientist,* September 18, 1986: 37.

107. Jack Fritz, "The Professionals' New Set of Tools," *South* Magazine, December 1986, vol. 74: 69.

108. Noel Berge, "Microcomputers in Development: Exploding the Myths," *Development International,* November/December 1986: 22–25.

109. Stan Aronoff, "Political Implications of Full Cost Recovery for Land Remote Sensing Systems," *Photogrammetric Engineering and Remote Sensing,* January 1985, vol. 51, no. 1: 41–45.

110. Ray Harris, "Contextual Classification Post-Processing of Landsat Data Using a Probabilistic Relaxation Model," *International Journal of Remote Sensing,* June 1985, vol. 6, No. 6: 847–866.

111. Arleen Cannata, "Casting a Wide Net," *Development International,* May/June 1987: 47–48.

112. Dr. Vishnu-Priya Sneller and Dr. Verly Jean-Francois, "Hardware, Software and Nowhere," *Development International,* May/June 1987: 14–15.

113. Marcus D. Ingle, Noel Berge, and Marcia Hamilton, *Microcomputers in Development: A Manager's Guide,* (West Hartford, Connecticut: Kumarian Press, revised edition 1986).

114. Lizbeth Shay, "Computers: Deleting Selection Errors," *Development International,* March/April 1987: 16–17.

115. The newsletters include: *Micro News,* from the Microcomputers in Planning Association, 10748 110th Street, S.E., Alto, Michigan 49302; the *DFD Newsletter,* from Data for Development, 122 Avenue de Hambourg, 13008 Marseille, France; the *ACCIS Newsletter,* from the Advisory Committee for the Co-ordination of Information Systems, Palias des Nations, CH-1211 Geneva 10, Switzerland; *Government Information Systems (GIS),* from the Economic Commission for Asia and the Pacific (ESCAP), United Nations Building, Rajdamnern Avenue, Bangkok 2, Thailand; the *Unisist Newsletter,* from the Division of the General Information Programme, Unesco, 7 Place de Fontenoy, 75700 Paris, France; and the *Development Communication Report (DCR),* from the Clearinghouse on Development Communications, 1255 23rd Street, N.W., Washington, D.C. 20037.

116. Gerald O. Barney and Sheryl Wilkins, *Managing a Nation: The Software Source Book* (Arlington, Virginia: Global Studies Center, 1986). The Center is at 1611 North Kent Street, Suite 600, Arlington, VA 22209.

117. Robert Schware, "Software Development in the Third World," *BOSTID Developments,* Board on Science and Technology for International Development, Winter 1987: 14–15.

118. Personal communication from Prof. R.S. Odingo, Department of Geography, University of Nairobi, Kenya, 1986.

119. Personal communication from Dr. Charles Paul, Manager, Remote Sensing, USAID, Washington, D.C., 1986.

120. Barber B. Conable, *Sound Ecology is Good Business* (Washington, D.C.: World Resources Institute, 1987).

121. Richard Gibson, "Chen seeks aid beyond IBM for supercomputer project," *Wall Street Journal,* January 13, 1988: 7.